topics in
ocean engineering

volume 2

volume 2

TOPICS IN OCEAN ENGINEERING

charles l. bretschneider, editor

chairman, department of ocean engineering
university of hawaii

gulf publishing company
houston, texas

Topics in Ocean Engineering

Library of Congress Catalog Card Number 78-87230

ISBN 0-87201-599-8

Topics in Ocean Engineering, Volume 2 is part of the University of Hawaii's Sea Grant Program sponsored by the National Science Foundation. It is based on a series of seminars given by experts in various disciplines of ocean engineering. The purpose of these seminars is to acquaint marine scientists, practicing engineers, laymen or technicians and graduate students involved with exploiting the sea with the "state of the art," which is changing rapidly because of the ever increasing knowledge of the marine environment and its effects on the prodedures and techniques being used to solve marine engineering problems.

We consider ocean engineering at the University of Hawaii in three broad categories: Port and Coastal Engineering, Offshore and Continental Shelf Engineering and Deep Ocean Engineering.

In *Topics in Ocean Engineering,* we attempt to cover the above broad categories. We also attempt to select experts from various parts of the United States, as well as from outside the country.

We do not intend to have the speakers necessarily present results of new research. Instead, we wish to make use of their knowledge and approaches they have gained in applying their own research to solving engineering problems peculiar to the marine environment.

The present series on *Topics in Ocean Engineering* includes speakers from the U.S., east and west coasts, one from Canada, one from Norway and several from the University of Hawaii.

For future series on *Topics in Ocean Engineering,* we intend to continue inviting speakers from various regions of the United States and other countries of the world.

Charles L. Bretschneider, Chairman
Department of Ocean Engineering
University of Hawaii

introduction

contents

topics in
ocean engineering

volume 2

li-san hwang and bernard
part 1: le méhauté

Harbor Design: Scale Model or Computer?

Dr. Li-San Hwang is manager of the Hydrodynamics Group at Tetra Tech, Inc., Pasadena, California. He has a B.S., M.S. and Ph.D. (hydraulics, hydrodynamics, physical oceanography).

Dr. Hwang received has B.S. at National Taiwan University, his M.S. in civil engineering at Michigan State University and his Ph.D. in civil engineering at the California Institute of Technology.

After graduation, he joined the National Engineering Science Company in Pasadena, where he became manager of the Hydrodynamics Department.

Since joining Tetra Tech, Dr. Hwang has researched and reported on many hydrodynamic and oceanographic phenomena. These include generation, propagation, runup and coastal effects of tsunamis. Currently, he is working on harbor response to periodic, dispersive or random incident waves; the effect of harbor agitation on ship moorings; and behavior of explosion generated waves.

Dr. Bernard Le Méhauté is vice president of Engineering Tetra Tech, Inc. in Pasadena, California. His educational credentials include a B.S., M.S. and D.Sc. (Hydrodynamics, coastal engineering, physical oceanography).

Dr. Le Méhauté received his doctoral degree with the highest distinction from the University of Grenoble. In 1952, joining the Neyrpic-Sogreah

Company in Grenoble, he spent 5 years as a coastal engineer. From Grenoble, Dr. Le Méhauté went to Canada to become professor at the Ecole Polytechnique in Montreal and later research professor at Queen's University in Ontario.

In 1961 he joined the National Engineering Science Company in Pasadena, California, and became director of the Geomarine Division. He was one of the founders of Tetra Tech and is now Vice President in charge of Engineering.

1

harbor design: scale model or computer?

Introduction

The problem of wave agitation in a harbor of complex shape has presented so many difficulties that engineers have relied almost entirely on scale model investigation. This method has, for a long time, proven successful for investigating the wave penetration in the range corresponding to the shortest component of incident wave spectrum; i.e., wind waves of wave periods smaller than 20 seconds.

In the case of incident long waves (wave periods longer than 30 seconds) the practical difficulties for reproducing satisfactory boundary conditions on scale model are such that the engineer has to carefully interpret the results of this method.

It will be shown how one can theoretically reproduce satisfactory boundary conditions on scale model, and how compromises have to be made for economic and practical considerations. A brief review of numerical methods of solution and a critical discussion on the limit of validity of these methods is given, together with a technical and an economical comparison between the experimental and numerical methods.

Harbor Problems

A harbor is intended to provide a protective environment for ships. It is clear, then, that harbor design is essentially related to the problem of ship motions and that we should begin by mentioning, in brief, what the ship problem is.

A freely floating vessel in a wave environment moves in a complicated fashion which, for convenience, is generally resolved into six components: heave, pitch and roll, surge, sway and yaw. Of these motions, the first group of three possesses spring-like restoration forces.

The second group of three motions, surge, sway and yaw, is not subject to such an effect if the vessel is free since no restoring force is present. Mooring the vessel provides the necessary restoration by means of the lines, either by weight or by elasticity so that again resonance is possible. It is such resonance in surge, sway or yaw with which one is most concerned since these are motions in the horizontal plane and thus promise, if large enough, collision with the surroundings. We shall not pursue ship motions any further except to mention that if the period is proper, very small amplitude waves are sufficient to cause damage.

Now it is pertinent to inquire just what the exciting waves are like. The waves arriving from deep water towards the coast represent a spectrum (Figure 1-1a) of frequencies from the very short to "sea and swell" caused by wind storms, either local or distant, to very long waves. Sea and swell may be thought to possess periods in the range from 4 to 20 seconds, while very long waves of interest may be typically of 1 to 3-minute periods.

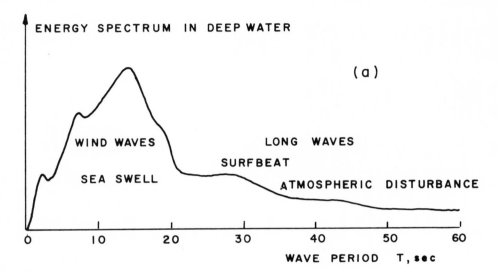

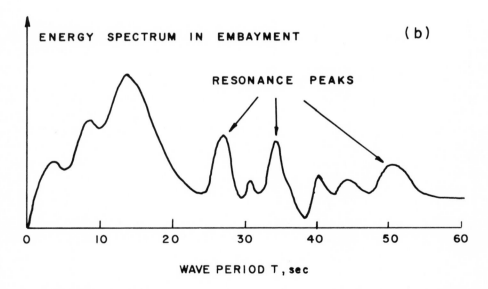

Figure 1-1. A comparison of typical wave energy spectrums in deep water and coastal regions indicates the influence of irregular bottom topography, refraction effects, multiple reflection and resonance effects in the range of long waves.

These ranges are specified to relate to two major classes of boats. Small pleasure boats of 20 to 50-ft. length are vulnerable to the shorter waves with periods, say, from 4 to 10 seconds. Large ships on the other hand, say, a 40,000-ton moored vessel may be vulnerable to waves in the 1 to 2-minute range.

Different design considerations must, of course, be made for the two types of waves as their behavior in a harbor are quite different. For brevity, we shall refer to the shorter waves as wind waves and to the long waves as "seiche"; although seiche is more properly the oscillation within the harbor caused by long waves.

In general, the statistical distribution of energy as a function of frequency at a given location is not very well known beyond the range of wind waves. It is reasonable to expect that the incident deep water energy spectrum is generally fairly flat between 30 to 200 seconds. This range of wave period triggers seiches in a harbor basin which are a cause of concern for large moored ships. However, due to peculiarities of refraction, reflection, multiple reflection and resonance over the continental slope and continental shelf, the shallow water energy spectrum presents a number of irregularities and peaks (Figure 1-1b).

As a matter of fact, physical behavior of long waves depends upon the bottom topography even in the deepest ocean. Moreover, due to small wave steepness (10^{-6}) the corresponding energy is not dissipated by breaking. The wave reflects back and forth from one continent to another. A relatively high energy level exists on the continental shelf caused by the reflective and refractive power of the continental slope. (Trapped modes, edge waves and bay resonance are of the same family of phenomena which are of particular significance in the case of long waves.)

In the case where preferential modes of oscillations of embayment and continental shelf match some of the dominant natural oscillations of a harbor basin, large seiche amplitude can be observed. Since one of the solutions to the seiche problem consists of mismatching peaks of harbor response curve with peaks of incident energy, it is of the utmost importance to determine the probability distribution of incident long wave energy spectrum.

Analysis of long wave records theoretically permits one to determine the probability of occurrence and intensity of the incident long wave energy. Such information rarely exists due to the length of recording time which would be required to obtain any significant results. Theoretically, a scale model or a numerical investigation which will reproduce the bottom topography up to the toe of the continental shelf should enable one to determine not only the response of a harbor basin but also the peculiarities of incident energy level at the entrance of the harbor. For practical purposes, one may prefer to build two scale models at two different scales. The first one at a small scale embracing a large part of the oceanographic topography which would provide the input for the second model reproducing the harbor and its immediate neighborhood only at a relatively larger scale.

In fact, such a theoretically ideal method is rarely accepted in engineering practice for obvious reasons of economy. The problem of seiche is simply solved by comparing various harbor layouts independently of the fact that some modes of oscillation may seldom be excited and others may have a very high probability of occurrence due to shelf and bay reinforcement.

Prior to any investigation on wave agitation in a harbor, one must have a reasonably accurate knowledge of the origin of the phenomena. For example, the scale model as well as the mathematical model will have to be completely different where one assumes that the seiche motion is due to long waves generated by atmospheric disturbances or is due to "surf beats." In the scale model, one has to generate long adjustable periodic waves while in the mathematical model, it may be necessary to reproduce short modulated wave trains of the wind wave range. Also, the scale model may be amenable to satisfactory linearized mathematical modeling. In the mathematical model, nonlinear interaction and shift of energy level from high frequency to low frequency defies, so

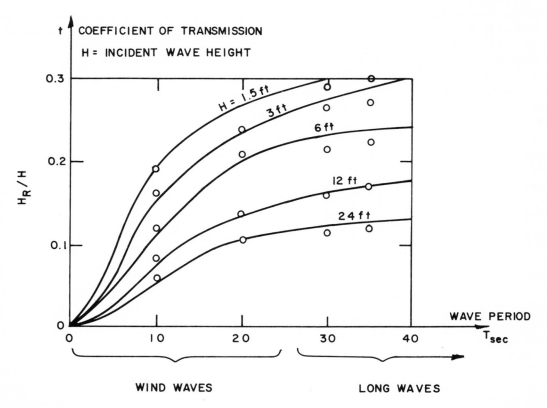

Figure 1-2. Typical rockfill breakwaters are permeable to long waves of small amplitude as evidenced by the value of the coefficient of transmission t vs. wave period T sec for various incident wave height H ft. More long wave energy may enter through breakwater than through harbor entrance.

far, mathematical analysis whereas it can be simulated on scale model.

Practical Solutions

The problem is to protect the vessels from waves by suitable harbor design. Any harbor can be considered as a resonant box, much like an organ pipe; it possesses natural modes of oscillation which, if excited by waves of corresponding frequency, will be induced to resonate.

The essential mechanism by which resonance is affected is partial reflection by boundaries. Con-sider a rectangular harbor of constant depth. An entering wave travels to the rear of the harbor where it is reflected by the wall and proceeds back toward the entrance. Upon reaching the entrance, part of the wave is reflected by the sudden enlargement while the rest escapes seaward. The reflected component remains within the harbor and repeats the cycle. Now if the frequency of arrival of waves is just right, this reflected wave combines with a new incident wave and amplifica-tion begins. The wave amplitude within the harbor continues to grow until the energy lost seaward at the entrance and frictional or breaking dissipation equals the arriving wave energy.[1]

In the case of wind waves, a great deal of protection is provided by simply preventing their arrival within the harbor through the construction of breakwaters in such a way that the arriving waves are reflected or largely absorbed. Design may be optimized to largely prevent the arrival of wind wave energy by a suitable choice of breakwater layout.

Of course, there must exist an entrance to the harbor and through this some energy will pass. This energy, as well as any surviving through the breakwater, can be effectively absorbed; for example, natural beaches are excellent wave absorbers because of energy dissipation due to wave breaking. Wave traps in rubble mounds are the most common and practical solution.[2]

It is more difficult to prevent the arrival of seiche within a harbor. The origin of seiche may be of several types. Seismic disturbance, large scale atmospheric phenomena, trapping of energy on the continental shelf, nonlinear interaction between wind waves corresponding to a shift of the energy level from high frequency waves to low frequency waves, surf beats, currents past the harbor entrance (generating alternating Karman vortices), even ship transit in and out of the harbor may induce seiche oscillation.

It is interesting to note that rockfill breakwaters are essentially transparent to very long waves (Figure 1-2[3]). For example, more wave energy of importance arrives within the Los Angeles Harbor through the breakwater than through the entrance to the harbor. Since breakwaters are transparent in both directions, harbor basins adjacent to breakwaters cannot trap any large amount of energy. So the problem of resonance only exists in inner basins, in which case attempts to limit the level of incident wave energy should be made. For this reason, rather be inclined to make a breakwater as impervious as possible to long waves despite the fact that seiche may now exist in offshore turning basins.[4] If breakwaters can be made essentially impervious, it may suffice to make internal basins, berths and quays "transparent" in order to avoid local amplification.

This can be done by building structures on piles through which the energy can easily pass rather than on solid bases between which oscillations may amplify. It is noted that wave absorption of very long waves is relatively inefficient. Long waves of small wave steepness do not break even on relatively gentle beaches and cannot be counted on to hold down the oscillation amplitude.[5]

More often one seeks to design the harbor such that whatever waves are unavoidable do not amplify. This can be done in three ways:

1. Mismatch the natural frequency of harbor oscillation with the largest spectral components of the incident waves. For example, in Figure 1-1b there are local peaks in the low frequency region which may be due to shelf oscillations. Designing the harbor so that its fundamental oscillation and first harmonics motion are removed from that frequency will suppress resonance.

2. Seek to mismatch the characteristics of successive basins. Figure 1-3a for example, shows a succession of basins in which the long wave agitation may increase from the sea to the inner basin if they happen to have the same natural period of oscillation. If the incident height is H and the individual basin amplification factor is A_i, the matched configuration allows growth to values $A_1 A_2 A_3 H$. On the other hand, making the natural period of one basin 50 seconds and the next 60 seconds gives a mismatch so that such growth will not occur (Figure 1-3b).

3. Manage to "leak" the trapped energy and thus eliminate resonance. In this, it is noted that a small amount of incident energy may cause large resonant oscillation so that only small amounts need be leaked to prevent growth. A practical method to accomplish this appears in Figure 1-4. Here two basins are connected by what may be a large pipe or a simple channel. Should resonance start in one, the excess energy is effectively drained so that growth is prevented.

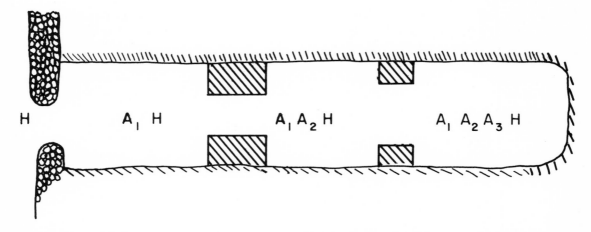

Figure 1-3a. The long wave agitation may increase from the sea to the inner basin when natural periods of oscillation of all basins match.

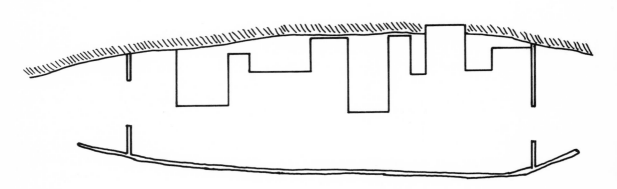

Figure 1-3b. Mismatching resonance periods or building platform on piles (wharf) limit long wave resonance effects.

Such are the types of solutions which the harbor designer may employ to prevent undue oscillation within a harbor. The methods by which the proper design may be found must now be discussed.

How To Find the Proper Solution

We shall not be primarily concerned with wind waves for the reasons already noted; they are shielded by even breakwaters permeable to long waves and they can be dissipated by wave absorb-ers. In studying seiche in a harbor, two main approaches are possible—experimental or analytical.

Scale Models

The oldest and still the most widely adopted technique of harbor design is the scale model study. The simplest way is to construct a model in geometric scale and to determine experimentally its characteristics. Then by trial and error, guided by experience, one may succeed in designing the best harbor layout.

If the seiche motion is due to surf beat, the choice of scale is the same as that which prevails in the case of penetration of wind waves; i.e., the model cannot be distorted and the scale is about 1:150. A smaller scale would cause too much scale effect by viscous and capillary action. The accurate reproduction of the generation process of surf beat reflecting towards the entrance of a harbor may require the reproduction of a large nearby bottom topography section.[6]

For seiche waves, large regions need to be modeled owing to the importance of shelf oscillations etc., and the model may need to cover a whole shelf section. The results of the first scale model can then be used as inputs for the second model which will be built to a larger scale, but covering a smaller area. If it is possible to treat a harbor separately, the first model can be built to a scale of 1:1000. Such a small scale, however, will usually require a distortion of the vertical scale so that excessively thin water layers will not result. A maximum permissible distortion is obtained by consideration of maintaining similitude of refraction. The general result is that the maximum depth which can be represented in a distorted model is inversely proportional to the square of the distortion; hence, the allowable distortion is limited by the maximum depth of the model and minimum wavelength.

Distortion of the depths, however, violates similitude of coastal reflection. It is then possible, and usually desirable, to correct this error by distorting the wave height. Distortion of depth increases reflection; an increase in wave steepness decreases reflection so that some compensation may be achieved.

A further practical problem relates to the satisfaction of boundary conditions. Clearly, a shelf region and harbor are considered as unbounded, being open to the sea. The walls of the test basin and the wave generating paddle are then an obvious problem since parasitic reflections from these surfaces always occur in such a way that instead of reproducing the characteristics of oscillation of bay and shelf, one has a tendency to excite the natural oscillations of the laboratory basin limited by the wall and paddle in which the model has been built.

It is necessary to install wave absorbers or filters around the periphery so as to avoid resonance of the whole model. As an order of magnitude, the dimension of absorbers and filters must be like one wavelength to be effective so that the model space allotted to these items can be necessarily great. Figure 1-5 shows an example of a typical configuration.

In the case of long wave oscillation induced by surf beats or nonlinear interaction, the wave filters

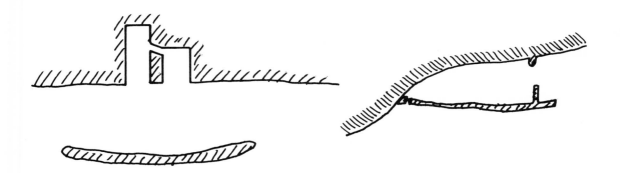

Figure 1-4. Providing a "leak," which avoids energy concentration, limits resonance.

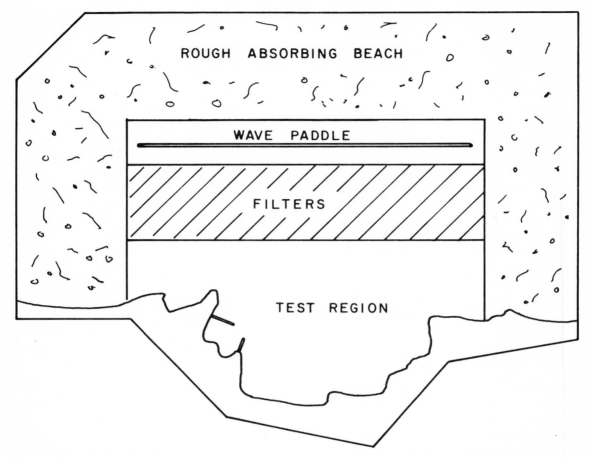

Figure 1-5. Scale model for studying long waves in a harbor would require reproducing the whole bay and shelf region, cumbersome wave absorbers and wave filters.

should be placed on the whole bottom of the model below the free surface. Short wind waves having a significant motion near the free surface only are seldom influenced by the filters and long wave components are damped as their motion is still significant in the deeper part of the model.

Finally, there are problems of adjustment of the wave characteristics. Wave direction, height and frequency must all be carefully controlled if the results are to be interpretable. Most sensitive is the adjustment of frequency since very small frequency change may cause large differences in the wave behavior inside the harbor. One is

therefore led to perform the tests by very slowly varying the frequency continuously so that no resonant jumps are missed. Were discrete frequencies used, it would be very easy to miss a sharp spike. This requirement of imperceptible frequency variation is responsible for long observation times and therefore higher costs.

Theoretical Approach

Constant Depth. Theoretical work in harbor response in the case of harbor layout of arbitrary shape is surprisingly recent, no doubt connected with the development of the large scale digital computer which makes prospects of success hopeful. We shall describe what is to our knowledge the most recent developments which are of practical use.[7,8]

One is dealing here with the governing equation:

$$\nabla^2 \Phi = 0 \qquad (1\text{-}1)$$

or Laplace's equation subject to boundary conditions. Velocity Φ is the potential. Considering for the moment constant depth, required boundary conditions are:

$$\eta = -\frac{1}{g} \Phi_t \qquad \text{at the surface}$$

$$\eta_t = \Phi_z \qquad \text{at the surface}$$

$$\frac{\partial \Phi}{\partial n} = 0 \qquad \text{at the coastal boundary}$$

$$\Phi_z = 0 \qquad \text{at the bottom}$$

$$\Phi = \Phi_0 \qquad \text{at infinity}$$

where η is the surface displacement. The third condition states that no flow occurs across the boundary; n is the unit normal at a boundary point and Φ_0 is the prescribed incoming wave from infinity. The problem configuration and coordinates are in Figure 1-6.

Since the water is assumed at constant depth, one is led to separate variables as follows:

$$\Phi = \frac{1}{i\omega} \phi(x,y) \, Z(z) \, e^{-i\omega t} \qquad (1\text{-}2)$$

where ω is the radian frequency. Application of the boundary conditions enables one to reduce the problem to the solution of Helmholtz's equation:

$$\nabla^2 \phi + k^2 \phi = 0 \qquad (1\text{-}3)$$

where k is the wave number. This solution is found by assuming the boundary to consist of sources of strength $Q(\xi, \eta)$ where ξ and η are coordinates along the boundary. Then one writes:

$$\phi = \phi_0 + \int_S Q(\xi, \eta) \, G(x,y: \xi, \eta) \, dS \qquad (1\text{-}4)$$

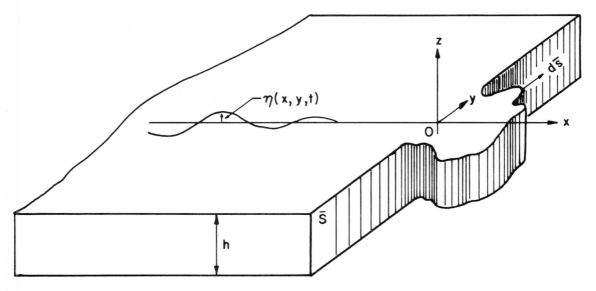

Figure 1-6. This schematic drawing shows a harbor with constant depth.

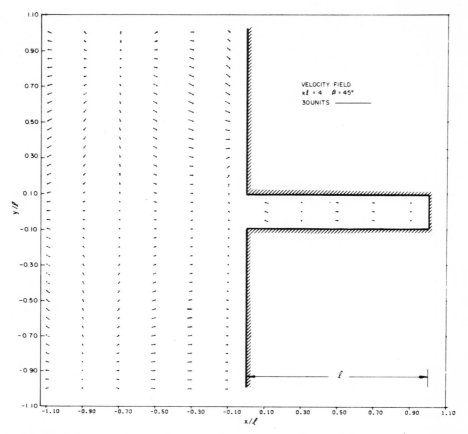

Figure 1-7. Amplification factor field in a rectangular harbor under the excitation of a wave arriving at 45° angle with the entrance. The asymmetry of the field is noticed.

That is, the ϕ at any point is the sum of the prescribed input ϕ_0 and the scattered contributions from the boundary. G is the appropriate influence function for the wave radiating from the sources:

$$G(x,y: \xi, \eta) = -\frac{i}{4} H_0^{(1)} (kR) \qquad (1\text{-}5)$$

where R is the distance between (x, y) and (ξ, η). The unknown is now Q, which must be determined. Application of the boundary condition

$$\frac{\partial \phi}{\partial n} = 0 \qquad (1\text{-}6)$$

gives[7]

$$\frac{1}{2} Q(\xi, \eta) + \int_S Q(\xi, \eta) \, G_n (kR) \, dS =$$

$$-\frac{\partial}{\partial n} \phi_0 (\xi, \eta) \qquad (1\text{-}7)$$

which must be evaluated numerically. This calculation is performed by dividing the boundary into short segments. The lengths of the segments must be such that several, say eight, are contained in one wavelength for accuracy. Within each segment, Q is assumed constant. The integral becomes a

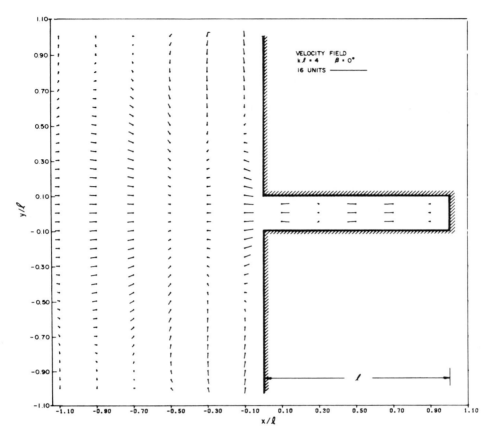

Figure 1-8. Instantaneous velocity field in a rectangular harbor under the influence of an incident wave arriving perpendicularly to the entrance. Location of node and antinode could be noticed.

summation, and one arrives at an algebraic rather than an integral equation:

$$\sum_{j=1}^{n} B_{ij} \, Q_j = b_i$$

where:

$$B_{ij} = \tfrac{1}{2} \delta_{ij} + \int_{\Delta S_j} G_n \, (X_j, Y_j; \xi_i, \eta_i) \, dS$$

$$b_i = - \frac{\partial}{\partial n} \phi_o \, (X_i, \, Y_i)$$

$$\delta_{ij} = \text{Kronecker delta} \qquad (1\text{-}8)$$

and $(X_j, \, Y_j)$ is the midpoint of segment ΔS_j. Ultimate solution involves inversion of the B matrix thus giving Q and hence ϕ and Φ.

This method has been programmed and has been applied to practical cases of harbors of arbitrary shape. In addition, programs can be written to plot the results in graph form. Such computer plots are shown in Figures 1-7 through 1-10 for rectangular and irregular harbors. Shown are plots of wave amplification as well as instantaneous velocity field.

Figure 1-11 is an aerial view of Barbers Point, Oahu Hawaii. Figure 1-12 shows the response curve at some locations of the harbor as function of the incident wave period.

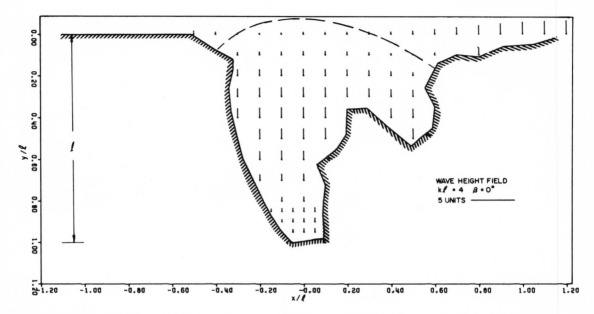

Figure 1-9. Amplification field in a bay of arbitrary shape assumed to be of constant depth. It is seen that the node cannot arbitrarily be located at the entrance.

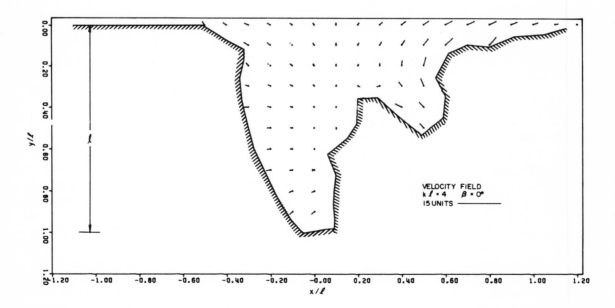

Figure 1-10. Instantaneous velocity field in a bay of arbitrary shape assumed to be of constant depth. It is seen that the smaller bay is subjected to resonant motion.

Figure 1-11. This aerial view is of Barbers Point Harbor (Oahu).

Variable Depth. The bottom topography of a bay or harbor usually is not even and the water depth varies from location to location. Thus, the theory described in the previous sections, which assumes a constant water depth may not be adequate due to the fact that the depth variation influences the wave amplification and resonance period of a bay or harbor. In order to take these effects into account, one has to solve the problem through the use of the equation which takes the depth variation into consideration.

As Figure 1-13 shows, the topography of a bay and its offshore area can be classified into three regimes: (a) shallow water region where variation of bottom geometry is large and usually has a significant effect on wave characteristics; (b) transitional region where bottom topography tends to be two-dimensional and the water depth is relatively deeper than in the shallow water region and small three-dimensional irregularities of the bottom topography can be neglected; and (c) the deep water region where bottom topography has little effect on the wave behavior. The geometry in the bay and near the bay is very complicated, so it is necessary to use numerical calculations to incorporate the bottom and shoreline effect variation. Such numerical calculations require a distribution of mesh points and must be sufficient

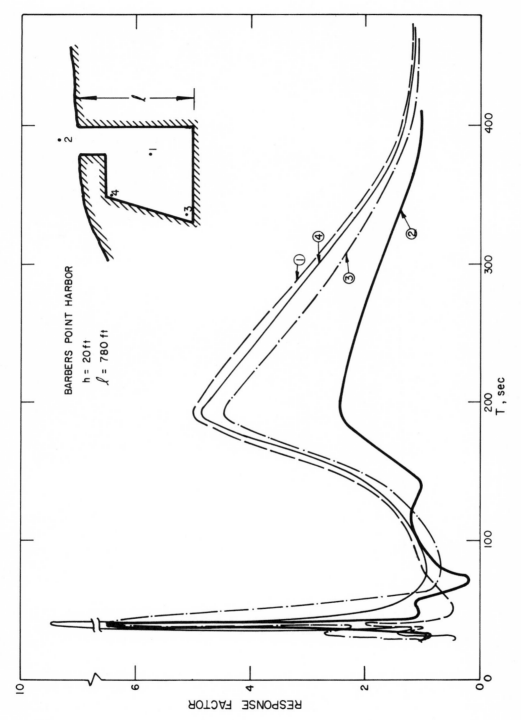

Figure 1-12. Response curves at four locations at Barbers Point Harbor. It is seen that the fundamental oscillation corresponds to the case where the basin is alternately filled and emptied with a practically horizontal free surface like a Helmholtz resonator in acoustic. Other peaks of the response curves correspond to initial oscillations.

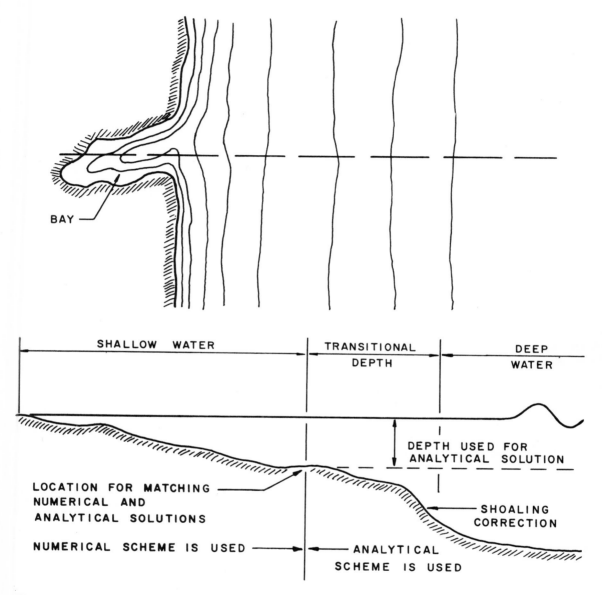

Figure 1-13. Different regions are classified corresponding to different mathematical treatment.

enough so that the variations of bottom topography and the wave behavior can be accurately represented. Limitation of the region of calculation must be small enough so that the calculation cost is feasible without affecting the accuracy of the results.

Since the behavior of waves inside the bay is related to incoming waves, the results obtained through the solution of the equation governing wave motion in the shallow water region without considering the nature of incoming waves is not sufficient. The outer region can be defined by an analytical solution obtained when considering the water depth as constant and equal to the depth at matching points (Figure 1-13). Then the analytical and numerical solutions are matched at the proper boundary. Using this scheme, incoming and reflected wave conditions will be properly considered.

Since the water depth used for the analytical solution is taken to be equal to the depth at the matching point, the amplification obtained inside the bay has to be modified by the effect of shoaling at the transition region; i.e., the incident wave height at the matching location has to be multiplied by a shoaling coefficient. With the introduction given above, we now proceed to outline the theoretical developments.

In the outer region where the water depth is assumed to be constant, the governing equation is the same as presented in the previous section. The boundary conditions on the free surface, at the bottom and at infinity are the same except at the boundary where the matching occurs. At the matching boundary, the wave characteristics inside the boundary have to be equal to the conditions outside the boundary.

The equation governing the wave motion in the outer region is Helmholtz's equation:

$$\nabla^2 \phi + k^2 \phi = 0$$

where ϕ is velocity kernel, which is related to the velocity potential by

$$\Phi = \frac{1}{i\omega} \phi(x, y) Z(z) e^{-i\omega t}$$

as it has been shown previously.

The solution of the outer region can be evaluated analytically by using the Green theorem through the contour integral around the outer half plane. If the matching boundary is chosen to be a straight line as Figure 1-12 shows, the condition on the matching boundary is simply

$$\phi(0, y) =$$
$$\phi - \frac{i}{2} \int_{-\infty}^{\infty} \frac{\partial \phi(0, \eta)}{\partial x} H_o^{(1)}(k \mid y - \eta \mid) d\eta \tag{1-9}$$

where η refers to the coordinate system along the matching boundary. This equation provides the link of the inner and outer region. The problem now is to solve the linearized long wave equation in the inner region with the above condition and the condition $\partial \phi / \partial n = 0$ along the shoreline.

In the inner region, the wave motion is influenced by the variation of bottom topography as has been discussed previously. The linearized long wave equation, which takes the bottom variation into consideration, should be used to describe the wave motion,

$$(h \phi_x)_x + (h \phi_y)_y = -\frac{\omega^2}{g} \phi \tag{1-10}$$

where   h is the water depth which is a function of x and y
   ω is the wave frequency
   g is gravitational acceleration.

The solution of this equation cannot be performed analytically as the water depth if a function of x, y. Furthermore, the shoreline around the bay may be quite complex. Therefore, the equation has to be solved numerically.

In order to minimize the numerical calculation, it is useful to introduce a fictituous boundary $\frac{\partial \phi}{\partial n} = 0$ at the locations;

   ac; db; x = 0,   y > a and x = 0   y < − b

as Figure 1-14 shows. The introduction of the fictitious boundary will not affect the results provided the distance of ab is sufficiently large in comparison to the size of the bay opening. Since the equation governing the matching boundary is singular, the numerical solution is a difficult one. A numerical method utilizing the numerical scheme[9] for the solution of linearized long wave

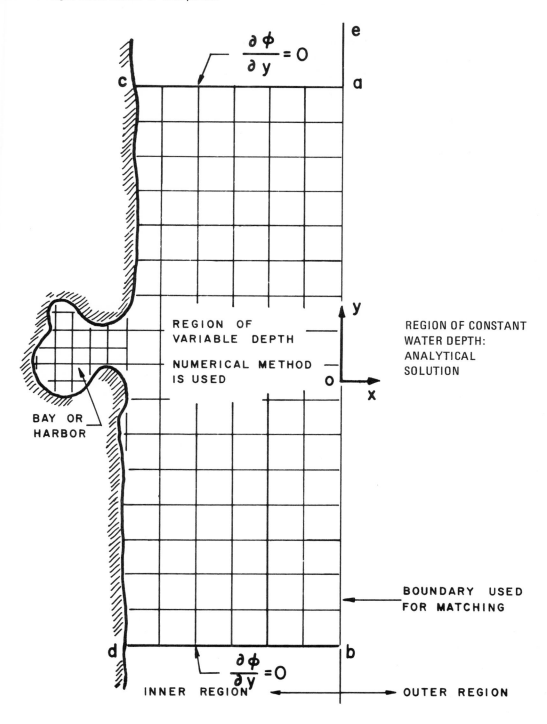

Figure 1-14. Schematic drawing is of a bay with reference coordinates.

equations has been developed, and it is found to be quite economical. Figure 1-15 gives sample results based on this method. It can be seen that the amplification factor is larger than the results obtained by the constant depth approximation.

Which Method?

One of the main limitations of both experimental and numerical methods has long been due to the difficulty of reproducing satisfactory boundary conditions on the sea side.

In the case of scale model studies, progressive wave filters[2] offer a satisfying solution theoretically even though it may be a relatively costly one.

Previously, all numerical methods require a priori the knowledge of the wave motion at a boundary from which the calculation proceeds. This motion can easily be measured in the case of tidal motion[10]. Otherwise, one has to arbitrarily assume the location of a node or an antinode[11]. In the latter case, the eigen-values for the bay limited by assumed boundary on the sea side can be determined, but the amplification factor cannot. The breakthrough in the two theoretical methods presented in the previous section is the *possibility of determining the response of a bay or a harbor to an incident wave without any assumption on conditions at the entrance as such conditions are part of the solution.*

Because of this advantage, it will be seen where the numerical method can now be considered, in some cases, a better method than a scale model study.

First, an analogy can be done between experimental and theoretical studies for the studies of long waves (Figure 1-16). The shelf region corresponding to variable bottom topography can be studied by mesh calculation, i.e., by the second numerical method; then the response of embayment can be determined. Since many harbor basins are dredged to constant depth, the first method based on a line source distribution is most suitable for studying the response of harbor basins to incident waves. Based on computing time, it is a more economical method.

Similarly, two scale models at different scales need to be built; the first one reproducing a large area of bottom topography proper to investigating the peaks of energy spectrum at the entrance of harbor basins. For evident cost reasons, it is not standard engineering practice to build a model of this kind before designing a harbor. The second model is built only limited to harbor basins and nearby topography. The harbor design is optimized without the necessary input which would indicate the peculiarities of the incident energy spectrum.

It will be of great importance then for a harbor designer to first investigate the range of applicability of all these methods as the sums saved can be significant. It is worthwhile to point out, for example, that the costs in computer time represented in Figures 1-7 through 1-11 are on the order of \$20 to \$50. (The boundary was defined by 60 point sources; i.e., approximately 8 point sources per wave length.) In the case of depth variation (second method), a similar result cost approximately \$100 of computing time in 1969 on a CDC 6600 computer. The construction of a scale model is at least \$3 square foot; a minimum cost of \$30,000 is necessary prior to obtaining the first experimental results. In addition, the results can be obtained by computer quasi instantaneously while experiments require a considerable length of time.

Methods Summary

Scale Models

1. The region to be modeled increases with increasing wavelengths while the horizontal scale used in the model decreases.
2. The boundary conditions are difficult to satisfy for long waves, easier for short wind waves.
3. The adjustment of wave characteristics is critical for seiche.
4. The nonlinear effects are reproduced in similitude. Those are of importance for waves of large steepness (wind waves).

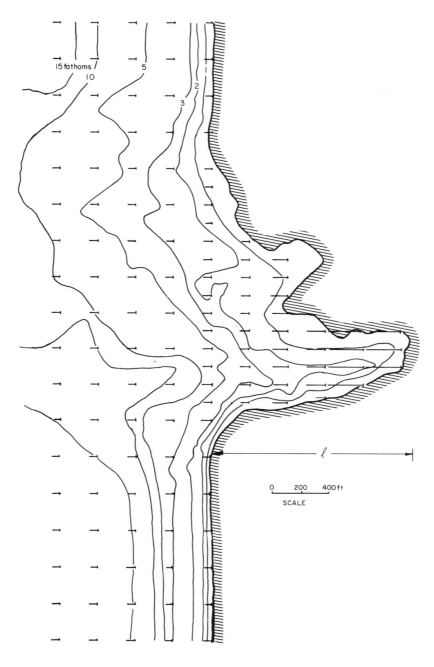

Figure 1-15. This illustrates an amplification field in a bay (Keauhou Bay, Oahu) of arbitrary bottom topography. The nondimensionalized incident wave number is $\frac{2\pi l}{L} = 0.4$.

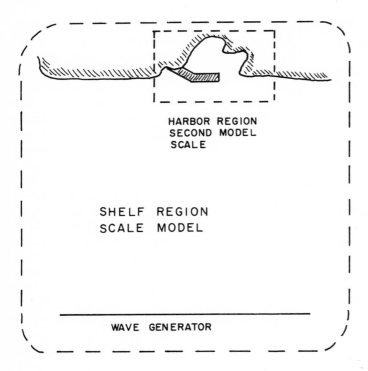

HARBOR REGION
SECOND MODEL
SCALE

SHELF REGION
SCALE MODEL

WAVE GENERATOR

Figure 1-16. A parallel between theory vs. computer follows: the first scale model should embrace the whole shelf region as the numberical method with depth variation. These two methods provide the most probable wave frequencies of high level energy at the entrance of the harbor. The second methods are aimed at studying the harbor layout.

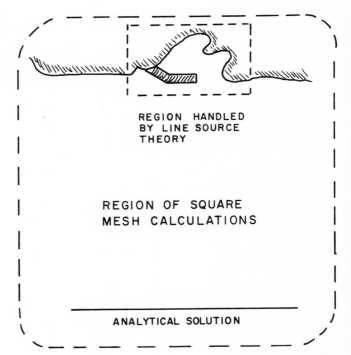

REGION HANDLED
BY LINE SOURCE
THEORY

REGION OF SQUARE
MESH CALCULATIONS

ANALYTICAL SOLUTION

5. Dissipative effects (wave absorbers) can be adjusted approximately in similitude based on the results of larger scale tests.

6. Permeability of breakwater to long waves can be adjusted by trial and error based on the results of larger scale tests.

Theoretical and Numerical Methods

1. Fine calculation intervals (segment length, grid size) are necessary for shorter waves; coarser for seiche.

2. The boundary conditions on the sea side can be reproduced easily.

3. The nonlinear effects are neglected which is valid for seiche of small amplitude.

4. Dissipative effects are neglected (which is valid for seiche of small amplitude).

5. Theoretical methods previously outlined can be refined to account for dissipation at the boundaries (beaches, wave absorbers) and breakwater permeability to long waves. The coefficients of reflection and transmission however must be empirical, i.e., based on large test experiments and/or past experience.

From these considerations it is seen that the theoretical approach tends to be more suitable to investigate long waves while scale model is a more reliable tool for studying the propagation of wind waves.

In brief, the theoretical problem is one of numerical computation; the computing cost increases with the cube of the number of line sources in the case of a harbor of constant depth, or the number of mesh points in the case of variable depth. Analytical study of short waves is costly since a very fine step-size (a fraction of a wavelength) is required for the numerical calculation. One is then limited by computer storage, size and speed. For very long waves, however, coarse intervals are sufficient and the price decreases. The theoretical method is particularly efficient in the case of relatively small size harbors. In large size harbors, the problem may have to be divided into separate problems between harbor basins for which interaction is negligible.

In the case of wind waves, computing costs, nonlinear effects and wave energy absorption permit the scale model to give a better return than the numerical approach. These considerations can be expressed graphically as in Figure 1-17. It is seen

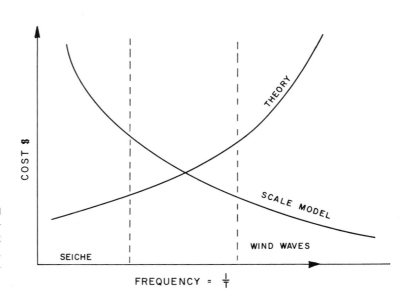

Figure 1-17. The computing time for studying the penetration of wind waves is such that scale model is the best method. Computer will more economically study seiche oscillation.

that the trends of costs for the two methods are opposite. A scale model for vanishing frequency requires modeling of every great oceanic region so that despite the decrease of scale the cost sky-rockets.

It is to be expected that future developments will favor theoretical approaches. Computers will become larger and faster so that the computations can be made more economically. Techniques based on theory or empirical results for handling absorption through breakwaters, breaking dissipation and so forth can be introduced easily in the computer programs.

In all, it appears that theoretical studies leading to numerical schemes and computerization with their attendant advantages should replace scale model studies of seiche. The coastal engineering community must be ready to exploit them effectively. Nonlinear effects and length of computing time still make scale model the best method for the study of wind wave protection in harbor basins.

Acknowledgment

The Atomic Energy Commission sponsored this work under contract number AT(26-1)289. The authors would like to express their appreciation for H. Lee Butler's contribution in numerical analysis and computer programming and David Divoky's assistance in preparing this paper.

References

1. Le Méhauté, B., "Theory of Wave Agitation in a Harbor," *Transactions, ASCE,* Part I, 127, Paper No. 3313 (1962), 364-383.

2. _____. "Wave Absorbers in Harbors," Contract Report No. 2-112 (1965), prepared for Waterways Experiment Station, Corps of Engineers under contract No. DA-22-079-civeng-64-81.

3. _____. "Permeability of Breakwater to Periodic Gravity Waves," ("Permeabilite des digues en enrochments aux ondes de gravity periodiques"), Thesis (University of Grenoble, 1957); *La Houille Blanche,* 12, No. 8 (1957), 903-931 and 13, Nos. 2 and 3 (1958), 148-179, 256-269, respectively, (translated by A.T.S., Inc.).

4. _____. "Harbor Paradox," *J. Waterways and Harbors,* ASCE proceedings, WW2 (1962), 173-185.

5. _____ ., R.C.Y. Koh, and Li-San Hwang. "A Synthesis on Wave Run-Up," *J. Waterways and Harbors, ASCE* (February 1968), 77-92.

6. Biesel, F. and B. Le Méhauté. "Notes on the Similitude of Small Scale Models for Studying Seiches in Harbors," *La Houille Blanche,* 10, No. 3 (1955), 392-407.

7. Hwang, Li-San and B. Le Méhauté. "On the Oscillations of Harbors of Arbitrary Shape," Prepared for the Atomic Energy Commission, Tetra Tech Report No. TC-123A, (1968).

8. _____ and E. O. Tuck. "On the Oscillations of Harbors of Arbitrary Shape," *J. Fluid Mech.,* (July 1970).

9. Stone, Herbert L. "Iterative Solution of Implicit Approximations of Multidimensional Partial Differential Equations," *SIAM J. Numerical Analysis,* 5, No. 3 (September 1968).

10. Leendertse, Jan J. "Aspects of a Computational Model for Long-Period Water-Wave Propagation," Rand Co. Report No RM-5294-PR (1967).

11. Wilson, B. W., J. A. Hendrickson and R. E. Kilmer. "Feasibility Study for a Surge-Action Model of Monterey Harbor, California," Waterways Experiment Station, Corps of Engineers under contract number DA-22-079-civeng-65-10 (1965).

part 2: e. eugene allmendinger

The Detailed Conceptual Design of the Oscilab

Eugene Allmendinger has been an associate professor of mechanical engineering and executive officer of the Office of Marine Science and Technology at the University of New Hampshire since 1961. He interrupted his work from 1968-69 to be visiting associate professor of ocean engineering at the University of Hawaii.

He has served as commissioner of the Maine/New Hampshire Advisory Committee on Oceanography. Professional organizations he belongs to include the Society of Naval Architects and Marine Engineers, American Society of Naval Enginners, Marine Technology Society and Sigma Xi (honorary research society).

From 1941-43 Mr. Allmendinger was a naval architect with the U.S. Maritime Commission; 1943-46, naval officer, U.S. Navy; 1946-49, naval architect, U.S. Navy; 1950-53, assistant professor, U.S. Naval Academy; 1953-58, assistant professor of naval architecture, MIT; 1958-59, associate professor of mechanical engineering, University of New Hampshire; 1959-61, professor of naval architecture, University of Sao Paulo, Brazil.

Author of 9 articles, he has also presented 4 major public lectures. Mr. Allmendinger received his B.S. in naval architecture from the University of Michigan in 1941 and his M.S. in mechanical engineering from the University of New Hampshire in 1950.

2

the detailed conceptual design of the oscilab

Introduction

The marine scientist has long dreamed of an ocean science laboratory that would allow him to move his laboratory into the sea and conduct research at the source of his interests for extended periods. The collective success of the U.S. Navy's Sealabs, Jacques Cousteau's Conshelf and other saturation-diving facilities has made this dream a realistic possibility. It has stimulated the imagination and interest of the oceanographic community and has produced a new breed of scientist, the "scientist-diver," who supplements professional expertise with qualification in saturation-diving techniques.

The U.S. Navy recognized the value of paralleling its Sealab programs with the development of a scientific sealab to support continental-shelf research activity. The interest in such a program was confirmed in a Bio-Dynamics, Inc. report, "Suggested Oceanographic Utilization of the Man-in-the-Sea Concept," published in September 1965. Consequently, the Office of Naval Research requested the University of New Hampshire Engineering Design and Analysis Laboratory (EDAL)

to conduct a detailed conceptual design study of a scientific sealab under the sponsorship of USN Contract NONR 3710-04. The results of this study were published in EDAL Report No. 100, "University Sealab." The report recommended and developed detailed conceptual designs for two scientific sealab systems. This paper discusses one of these systems, the Ocean Science Laboratory (Oscilab).

University Sealab Study

The objective of the University Sealab study was to develop the detailed conceptual design of a scientific sealab that would meet the immediate needs of the scientist-diver. This design was to be within the boundaries of existing technology and in sufficient detail to permit reasonably accurate estimates of capital and operating costs.

The study was conducted in three phases. Phase 1 called for determining the requirements for the scientist-diver in a saturation-diving facility. In phase 2 the most feasible saturation-diving system based on various overall system concepts was selected. In phase 3 a detailed conceptual design of the chosen system was developed.

Two systems were selected—the Oscilab and Seadopod systems. The Oscilab system is "bottom-oriented": the laboratory-habitat remains at its submerged location for the duration of the mission. The Seadopod system is "surface-oriented": a pressurized laboratory-habitat is located on a surface vessel, and the scientist-diver invades the ocean depths for short periods in the Seadopod, which is suspended from the surface vessel.

Groundwork for the Oscilab Design

The detailed conceptual designs of the Oscilab and the Seadopod were based on two sets of considerations: the requirements of the scientist-diver and the designer's technical criteria.

Determining the requirements of the scientist-diver involved the following:

1. Surveys of oceanographic research studies made in conjunction with projects such as Sealab II, Sealab III as projected, Conshelf and Westinghouse Corporation's Man-in-the-Sea activities;
2. Research proposals prepared specifically for this project by scientist-divers from several academic institutions planning to use a scientific sealab;
3. Reviews and analyses of these requirements by the EDAL team;
4. Reviews of these requirements with the Scientific Advisory Committee.

The requirements were established as follows:

1. *Site-to-site mobility:* The facility must be easily transported over the ocean at a minimum speed of 3 to 5 knots and must be capable of supporting research activity within 24 hours of arrival at the site.
2. *Horizontal mobility:* Diver-scientists must be capable of working within a 300-ft radius.
3. *Depth of research activity:* The facility must be capable of supporting divers in depths of 50 to 300 feet.

4. *Personnel at saturation pressures:* The proposed research will require three to four diver-scientists and two supporting divers.
5. *Disturbance to environment:* Disturbance must be held to a minimum. There must be no effluent and waste from the facility.
6. *Maximum duration of personnel at saturation pressures:* This is two weeks.
7. *Depth flexibility on site:* It is desirable, but not essential, to be able to alter the depth of the diver-support facility at a research site easily and facilitate work at two or more depths at one site.
8. *Need for personnel to remain on bottom continuously:* This is not absolutely necessary.
9. *Need for special visibility ports* (inside a habitat to outside): This is not essential because visibility is limited in most research locations. Simple visibility ports are desirable.
10. *Maximum sea state for routine operation:* Five is maximum.
11. *Environmental conditions for research studies:* They are

    Temperature—32-85°F
    Depth—50-300 ft
    Bottom composition—Rocky to oozy
    Bottom topography—Flat to steep
    Visibility range—0 to clear
    Currents—0-1 knot for research
    　　　　　0-5 knots for habitat integrity

12. *Research bench space:* This amounts to 20-in benches with a total of 3 ft for each man in the habitat.
13. *Research instruments and electronic systems space requirements:* A minimum of 50 ft^3 is required.
14. *Research supplies and sample storage:* A minimum of 50 ft^3 with part of the sample storage in a deep freezer, is required.

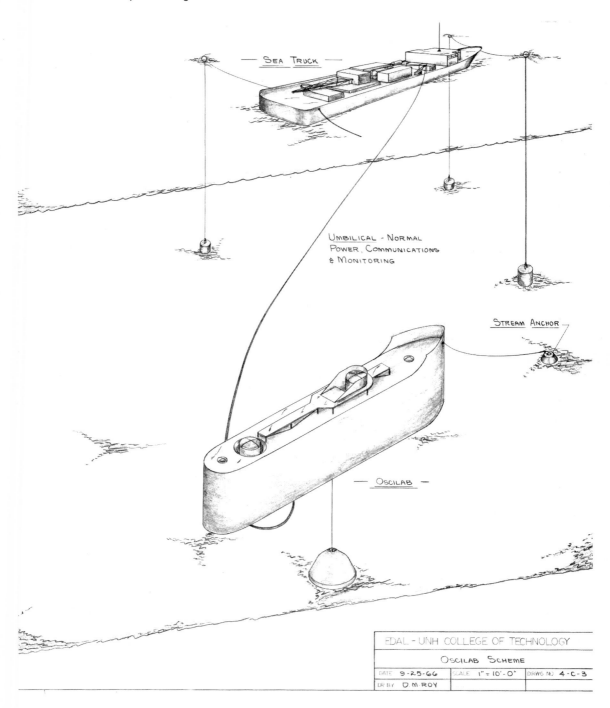

SEA TRUCK

UMBILICAL - NORMAL
POWER, COMMUNICATIONS
& MONITORING

STREAM ANCHOR

OSCILAB

EDAL - UNH COLLEGE OF TECHNOLOGY

OSCILAB SCHEME

| DATE 9-25-66 | SCALE 1" = 10'-0" | DRWG NO 4-C-3 |
| DR BY D M ROY | | |

Figure 2-1. Oscilab System Components.

15. *Environmental monitoring:* As a minimum, the parameters necessary on a continuous basis include
ocean temperatures at the bottom;
dynamic pressures of the ocean bottom;
salinity of the ocean near the bottom;
bottom currents; and
ambient light levels at the bottom.
It is assumed that the habitat environment will be monitored on a continuous basis and in a form that can be used as research data.

16. *Diver location:* A system is required to locate the diver relative to a known bottom fix with an accuracy of about one meter.

17. *Site survey:* A site survey must be conducted. If the facility cannot do this, a research submersible will be required to conduct this task.

The designer's technical criteria assembled by the EDAL Team include the following:

State-of-the-art design philosophy
Safety
Operational reliability
Decompression flexibility
Fiscal considerations, capital and operating
Habitat stability
System simplicity and ease of operation
Independence of adverse seas and weather
Habitability
Personnel training requirements

Oscilab Design Concept

This section describes the Oscilab saturation-diving facility design concept in two parts, the Oscilab system components and the Oscilab system operational concept phases. The overall Oscilab system is shown in Figure 2-1.

Oscilab System Components

The Oscilab system components include the laboratory-habitat, the surface-support ship, and the surface-subsurface linkages. These components are described in the following paragraphs.

Laboratory-habitat

The laboratory-habitat (the Oscilab) is a nonpropelled vessel designed to permit six men to live and work continuously at a maximum depth of 300 ft for a two-week period under saturation-diving conditions. The Oscilab has the following principal geometric and hydrostatic characteristics:

Length	71.0	ft
Breadth	10.0	ft
Depth	18.0	ft
Displacement—Surface, main anchor dry (salt water at 64 lb/ft^3)	174.75	LT
Displacement—Submerged (salt water at 64 lb/ft^3)	202.65	LT
Centers of Buoyancy (reference—amidships and keel)		
Surface:		
Vertical	9.3	ft
Longitudinal	0.0	ft aft
Submerged:		
Vertical	10.4	ft
Longitudinal	0.1	ft aft
Centers of Gravity (reference—amidships and keel)		
Surface:		
Vertical	6.7	ft
Longitudinal	0.1	ft aft
Submerged:		
Vertical	6.3	ft
Longitudinal	0.1	ft aft
Stability Data (uncorrected for free surface)		
Surface:		
GM (Transverse)	3.0	ft
GM (Longitudinal)	32.3	ft

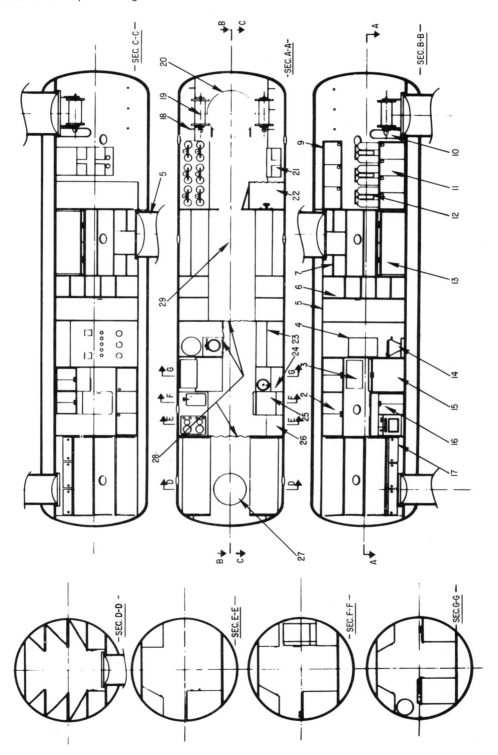

Figure 2-2. Oscilab Interior Arrangement.

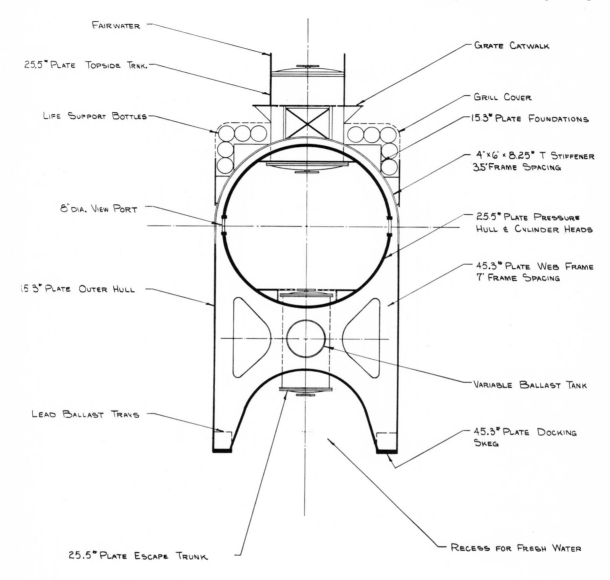

Figure 2-3. Oscilab Midship Section.

Submerged:
  BG   4.1   ft

It should be noted that the preceding characteristics differ from those for the Oscilab in the University Sealab study. These revisions were the result of design refinements made in conjunction
with the succeeding Oscilab model study program.

The 40-ft cylindrical pressure hull is 9 ft in diameter, such dimensions being dictated by the
availability of this major structural item. The ends
of the cylinder are closed by elliptical heads. The

interior arrangement from forward aft, as shown in Figure 2-2, includes the wet room, laboratory, vessel control space, and general living area. Three access trunks penetrate the hull: a 48-in diameter wet room trunk; a 30-in diameter escape trunk in the living quarters; and a 48-in diameter trunk in the laboratory space which provides topside access and also serves as a lock. Other hull penetrations include those for viewports and electrical, mechanical, and piping leads. With the exception of the viewports and topside access trunk, all penetrations are located as near the bottom of the hull as possible to minimize the dangers of flooding and loss of breathing gas in the event of a hull-fitting failure.

Principal components of the exterior hull structure for the pressure hull, shown in Figure 2-3, are seven frame ring stiffeners and six "saddle" web frames. This structural arrangement permits exterior as well as interior loadings of the pressure hull and allows the use of a thin hull plating. Heavy longitudinal flat bars or keels are fitted at the bottom of the web frames, port and starboard, for dry-docking; and lead ballast trays are located on these flat bars.

The ends of the vessel are transversely framed. Twin vertical keels support the end ballast and buoyancy tanks, the battery, and the personnel transfer capsule (PTC) well structure.

An outer hull of light plating forms a fairing and protecting envelope for the pressure hull, external tankage, and equipment. The hull is "wall-sided" in way of the cylinder, but some "ship-shapeness" is provided at the ends to reduce towing resistance and enhance the Oscilab's sea-keeping characteristics. A small weather deck, protected by a bow spray-shield and access-trunk fairwater, facilitates surface handling and servicing procedures.

The equipment and tankage outside the pressure hull are shown in Figure 2-4. The forward and after ends of the vessel are occupied by the main ballast and buoyancy tanks, the battery tank, and the PTC well. The super structure houses the main-anchor winch, the stream-anchor winch and twenty breathing-gas bottles, which are 20 ft by 9

5/8 inches. The space below the cylinder houses the fresh water bags, the variable ballast tanks, and the main anchor recess. Flasks containing high pressure air for blowing the main ballast tanks are in the buoyancy tanks.

The main anchor system and anchor are shown in Figures 2-5 and 2-6. This system serves two functions: it provides the motivating force for the "interface breakthrough" and provides the bottom anchor for winching-down and winching-up. The anchor has a lower and an upper section. The lower section has the ballast capacity to furnish the "interface breakthrough" force and is flooded before the anchor is lowered from the surface. The upper, much larger section furnishes the main anchoring weight and is flooded when the anchor bottoms. Both sections are blown while the anchor is on the bottom. Upon retrieval, the anchor is drawn into the rain anchor recess where it is completely housed by the Oscilab hull. (The main anchor system is described in detail later.)

Surface Support Ship

For design analysis, the surface support ship is a "Sea Truck" (or similar vessel), and its principal characteristics are

Length	138	ft
Beam	29	ft
Draft (loaded)	about 7.5	ft
Clear deck area	102 x 29	ft

The ship's deck houses are confined to roughly the forward quarter length. The remaining three-quarters of the length is a flat working deck. This feature makes Sea Truck an extremely versatile support ship, since its deck load can be composed of a variety of modular arrangements—each arrangement meeting the requirements of a specific mission.

The design concept of Sea Truck also reduces port time and nonproductive port expenses to a minimum. Modular units (such as vans, decompression chambers and power units) can be prepared ashore before the ship's arrival. Then the units

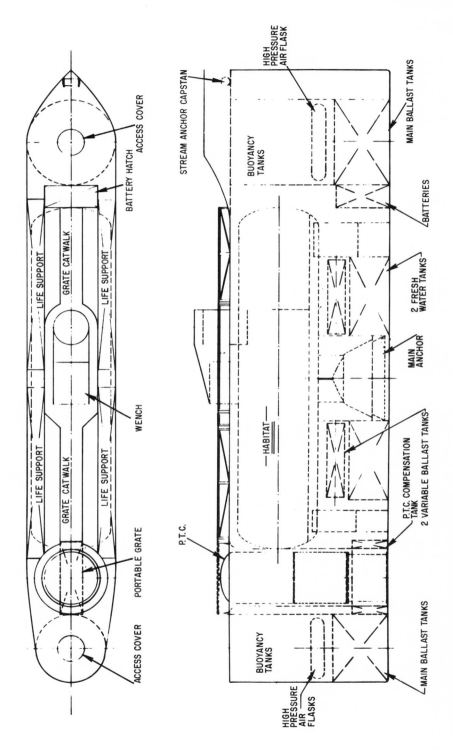

Figure 2-4. Oscilab Propile and Deck Plan.

must only be hoisted aboard and secured to the deck. Units of heavy, concentrated deck loads may require local (and temporary) reinforcement of the ship's structure.

The Oscilab system requires that Sea Truck carry the following:

1. Two Diesel generator sets, one for stand-by
2. One reel and associated equipment for tending the umbilical
3. One decompression chamber
4. One deck crane for handling the PTC

5. Communications and monitoring equipment (in a van)
6. In-transit accommodations for divers and accommodations for any specialized surface-support personnel that may be required
7. One 25- to 40-ft power work boat
8. Miscellaneous equipment required to support the mission (in a van)

Essentially, the Sea Truck is required to furnish the Oscilab with power under normal operating

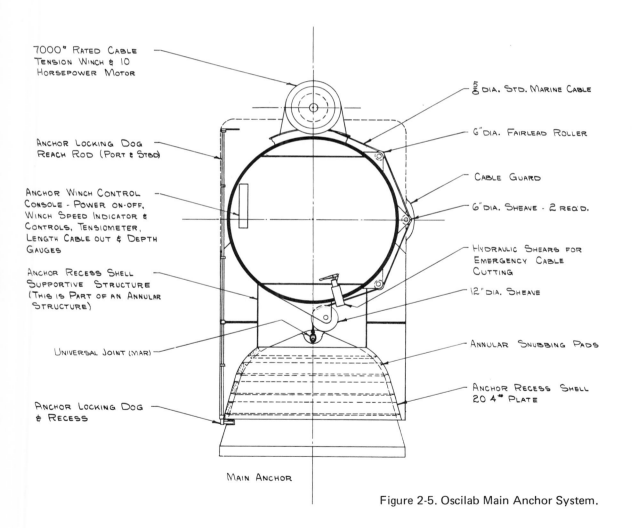

Figure 2-5. Oscilab Main Anchor System.

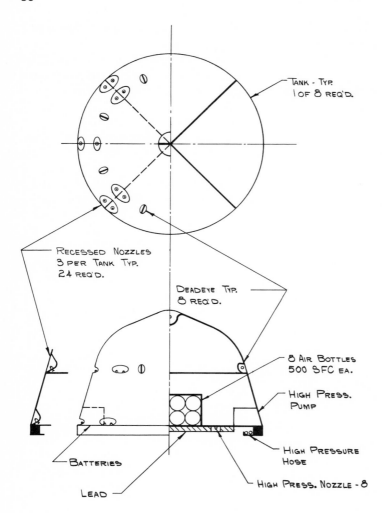

Figure 2-6. Oscilab Main Anchor.

conditions as well as provide emergency recompression and decompression facilities.

Surface-subsurface Linkages

The surface-subsurface linkages in the Oscilab system include a personnel transfer capsule (PTC) and an "umbilical" cable.

The PTC, shown in Figure 2-7, is carried in a well just aft of the Oscilab pressure hull. It is entirely independent of the Oscilab and provides for the emergency refuge and escape of the six divers. In other words, it serves as the Oscilab's "lifeboat." The PTC's life-support system will sustain six men for six hours. Controlled surfacing is achieved by the PTC's winch that pays out the downhaul cable attached to the Oscilab's keel at a specific rate. The PTC winch and lead ballast are located on a "raft" about 4 ft below the PTC entrance trunk. If the winch system fails, the raft can be released from the PTC. The raft is detached before mating the PTC and the deck decompression chamber.

The PTC is designed to have positive buoyancy in the full-load condition. Hence, no winch drive or power source is required if the capsule is used exclusively for one-way escape. It has been estimated that a return to the Oscilab by the PTC (a winching-down procedure similar to that employed by the McCann submarine rescue chamber) would require a 3-horsepower motor and a 1200-watt-hour battery for 30 minutes of operating time. The umbilical contains cables for normal power supply, communications, visual monitoring, and breathing-gas monitoring and is stowed on a reel on the support ship where it is controlled.

Oscilab System Operational Phases

The Oscilab system's operation concept is presented schematically in Figures 2-8 and 2-9 and in the following discussion, asterisks indicate the phases shown in the figures. In the discussion of each phase the divers, the Oscilab, and the support ship are considered unless those elements are not germane.

Preparation for Sea

Divers. Six divers are selected; four diver-scientists to conduct the research programs and two diver-crewmen. All divers must be thoroughly trained in saturation-diving techniques. In addition, they must be completely familiar with the

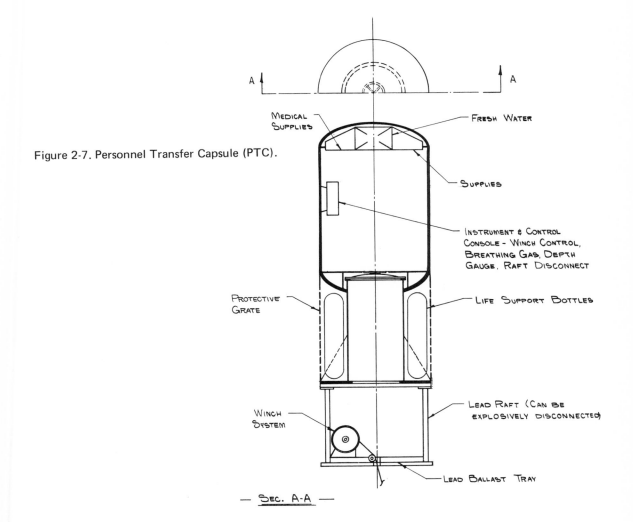

Figure 2-7. Personnel Transfer Capsule (PTC).

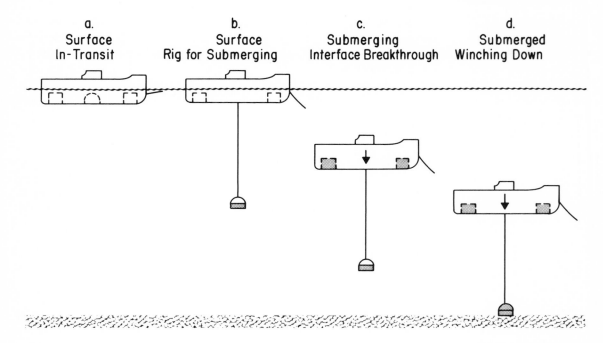

| a.
Surface
In-Transit | b.
Surface
Rig for Submerging | c.
Submerging
Interface Breakthrough | d.
Submerged
Winching Down |

Figure 2-8. Oscilab Operating Phases.

general aspects of the Oscilab's operation and the details of emergency procedures. The two diver-crewmen must be thoroughly trained in all phases of the Oscilab's operation since this will be their primary duty. They should also be familiar with the broader aspects of the research to be accomplished so that they may effectively aid the diver-scientists. One of the diver-scientists is designated "chief scientist," and one of the diver-crewmen "captain." Regarding safety and conduct of the mission, the captain makes the final decision, but he is expected to be flexible. All major operational decisions should be cleared with the proper authority on the support ship if conditions permit.

Oscilab. The Oscilab is made ready for sea in all respects including the testing of all systems. To ensure safe and proper operation, the following procedure must be strictly observed in loading all items:

1. All items must be stowed in their assigned locations.

2. A record must be kept of the weights of all items.

3. Weights must be kept within the design weight allowance.

4. No item can be brought aboard which represents a potential hazard when exposed to a helium-oxygen atmosphere.

5. When the Oscilab is fully loaded, the fore and aft draft must be carefully recorded by draft gauges and the weight and location of variable ballast water adjusted to place the Oscilab at its submerging trim waterline. The six divers are to be aboard and at their submerging stations for this procedure.

It should be emphasized that the variable ballast system is designed to permit flexibility in load weight and longitudinal distribution. A total loading condition outside the compensating capacity of the variable ballast system requires a weight and moment adjustment in the lead ballast. This ad-

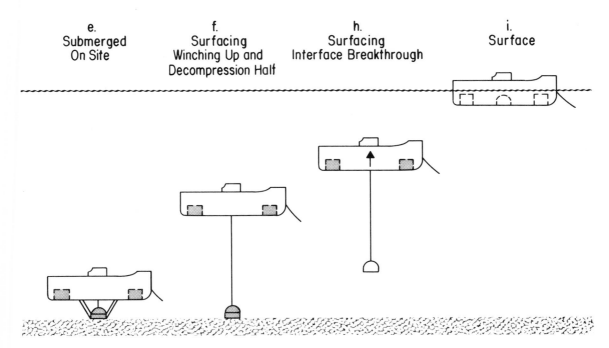

Figure 2-9. Oscilab Operating Phases.

justment can be made to a limited extent. However, the lead is located at the keel, and any lead removed to compensate for an added load weight will always decrease the stability. Conversely, a removal of load weight and an addition of lead will increase the stability.

The Oscilab is given an internal pressure test of 190 psi (using air) just before departure.

Support Ship. All modular units are placed aboard and secured to the Sea Truck's deck. The umbilical is temporarily connected to the Oscilab and all surface-subsurface systems are thoroughly tested. All other supplies and equipment required for the Sea Truck's operation are taken aboard.

Surface—In Transit

Divers. The divers may live aboard the support ship during transit, or if the transit time is long, they may be flown to the site by helicopter.

Oscilab. The Oscilab is unoccupied during transit. All hatches and hull stop valves are closed with watertight covers fitted over viewports and cable entry tubes, and the umbilical is not connected. Proper navigation lights are powered and controlled from the support ship. Battery powered emergency lights are also available. The main anchor is dry and housed within the outer hull.

The pressure hull is charged with 50 psi of air; this pressure is monitored on the support ship for indications of leakage. A compressed air line, suspended from the towing cable, supplies make-up air to the Oscilab if there is a pressure drop. Suspected leakages are checked and, if possible, stopped by scuba divers diving from the support ship's work boat. If feasible, a shallow water route to the site should be followed.

Support Ship. In addition to towing, the support ship's primary purpose and function is to monitor the Oscilab's internal air pressure, thereby checking for signs of leakage.

Surface—Rig for Submerging

Divers. In preparation for submerging, the divers board the Oscilab and make a "dry" entry through the topside access trunk after the interior pressure has been reduced to 1 atmosphere. They then go to their submerging stations. The two diver-crewmen will be in the control space. Two diver-scientists will be aft in the living space and the other two diver-scientists in the wet room.

Oscilab. The "rig-for-submerging" procedure is as follows:

1. Attach the umbilical and test all surface-subsurface systems.
2. Lower the bow stream-anchor to the bottom. The Oscilab falls off on the stream anchor until proper scope has been obtained.
3. Adjust the variable ballast weight to compensate for a departure from the "standard" specific weight of seawater, 64 lb/ft³. (Specific weights of seawater at several depths in the water column at site should have been previously obtained. This procedure is necessary because seawater varies in specific weight from about 63.8 to 64.3 lb/ft³. This variance could cause the Oscilab's submerged displacement to vary as much as 3,000 pounds. The variable ballast system is designed to compensate for this change from "design" displacement as well as for changes from the design load weight.)
4. Cast loose the towing cable. This is done topside by a crew member of the support ship.
5. Flood the lower part of the main anchor and lower it to within 40 ft of the ocean bottom.
6. Close the topside trunk hatch, open the air-vent line, and start charging the Oscilab with helium. The Krasberg pO_2 meter will now feed the required amount of oxygen into the atmosphere. The Oscilab is pressurized to site-depth pressure.

Support Ship. Before the Oscilab's submergence, the support ship retrieves the towing cable and lays out a three or four-point moor on a course approximately parallel to the Oscilab's position and at a stand-off distance of 150 feet.

Submerging—Interface Breakthrough

"Interface" is the depth range 0 to 50 ft. This is the anticipated depth range for the severest surface wave action that could affect relative motion of the Oscilab with respect to the ocean bottom. In a seaway, a cable connecting a bottomed anchor to the Oscilab in the interface would be subject to excessive snap stresses and the possibility of kinking and breaking. In this design concept, the main anchor is lowered to within 50 ft of the bottom. Thus cable tension is maintained until the Oscilab is submerged to a depth of 50 ft and the anchor is bottomed. At this depth, there is very little relative motion between the anchor and the Oscilab; commensurately low dynamic stresses are placed on the cable.

Divers. All divers are at their submerging stations during the interface breakthrough. The diver-crewmen, stationed at the controls, vent the main ballast tanks. They anticipate the bottoming of the main anchor by activating the winch (at a slow speed) at a depth of 35 ft, thus minimizing cable slack due to inertia forces. When tensiometer readings show that the anchor is bottomed, the winch is stopped. Steady depth and tensiometer readings indicate that the system is in static equilibrium at 50 feet.

Oscilab. A description of the Oscilab's operation during interface breakthrough will be facilitated by reference to fundamental equations in which

V_{MB} = Volume of the main ballast tanks; its magnitude is chosen to provide adequate freeboard and seaworthiness characteristics when the Oscilab is fully surfaced

V_{LA} = Volume of lower part of main anchor

V_{RB} = Volume of reserve buoyancy above the submerging trim waterline

S = Specific weight of sea water

F = Net downward force, a basic parameter chosen to provide the Oscilab with a suitable average interface-breakthrough velocity

D_S = Submerged displacement force (not including main anchor)

W = Oscilab "built" weight plus weight of full load

W_S = Oscilab submerged weight

C_M = Anchored cable tension, moored

C_D = Anchor cable tension, dynamic

While descending through the interface, the following expressions are obtained:

$$V_{MB} < V_{RB}$$

$$V_{MB} + V_{LA} > V_{RB}$$

$$V_{MB} + V_{LA} - V_{RB} = V$$

$$\therefore F = SV \qquad (2\text{-}1)$$

When the anchor bottoms, the anchor weight is removed from the system and the following equations are obtained:

$$D_S{}' = W + W_A + SV_{RB}$$

$$W_S = W + SV_{MB}$$

$$SV_{MB} < SV_{RB}$$

$$\therefore D_S > W_S \qquad (2\text{-}2)$$

The moored cable tension required for static equilibrium is

$$C_M = D_S - W_S \qquad (2\text{-}3)$$

The cable tension during descent, discussed in the winching-down phase, is

$$C_D = C_M +$$

hull drag at the winching-down velocity $\qquad (2\text{-}4)$

The limiting value of C_D is determined by winch and cable load capacities. For this design, C_D is about 7,000 pounds.

When the main anchor bottoms, the Oscilab regains positive buoyancy with no action by the diver-crewmen required.

Support Ship. The support ship's primary function during this phase involves the controlled reeling-out of the umbilical.

Submerging—Winching Down

Divers. During the winching-down phase, all divers are at their submerging stations. The diver-crewmen once again check all surface-subsurface systems and report the results as well as the general conditions in the Oscilab to the support ship. If conditions are satisfactory, the diver-crewmen activate the main-anchor winch, carefully noting readings on the depth gauge, the tensiometer, and the internal-external pressure gauge. (The internal-external pressure gauge is a special instrument designed to record the Oscilab's atmospheric pressure, the sea pressure at the Oscilab's depth and the pressure differential.)

Oscilab. The Oscilab's descent is now due entirely to the winching-down force of the cable, C_D. As noted earlier,

$$C_D = C_M + \text{hull drag at winching-up velocity}$$

The winching-down velocity is primarily a function of the load capacities of the winch and cable.

The submerged displacement, D_S, is the product of the submerged volume of displacement and the specific weight of seawater. The change in the submerged volume, due to hull compressibility, will be slight and may be neglected. However, the specific weight of the Oscilab is primarily a function of seawater temperature and salinity; it will therefore most likely increase, causing a small increase in the submerged displacement and in the cable tension, C_D. This increase, of course, will be recorded on the tensiometer and may be compensated for by admitting a small amount of water to the variable ballast tank if necessary.

Support Ship. The support ship continues with the controlled reeling-out of the umbilical.

Submerged—On Site

Divers. When the Oscilab is on site, the divers check all surface-subsurface systems and scan the bottom and main anchor by television. When the pressure is equalized, the wet-room hatch is opened and one of the diver-crew makes the first excursion dive to examine the anchor and to connect the Oscilab to the anchor by preventer bars. The on-site routine can now begin.

On-site watch and duty bills are established. One member of the diver-crew remains aboard the Oscilab at all times for monitoring and to maintain communications with the surface ship and the divers working outside. The diver-scientists make excursions in pairs, accompanied by a diver-crewman if necessary.

A careful record must be kept of all on-site weight changes, such as changes incurred by bringing heavy specimens aboard and by the depletion of supplies. Adjustment in variable ballast is made to maintain approximately a zero load change. Load changes are checked by monitoring the cable tensiometer after the preventer bars have been removed just before ascent.

Oscilab. The longitudinal attitude (trim) of the Oscilab is independent of the ocean bottom topography. The main anchor may be inclined, but the Oscilab, tethered to the anchor and possessing a large amount of submerged stability, should have little or no trim. Preventer bars are secured between the Oscilab and anchor as cable back-up devices.

Support Ship. The support ship continues to supply normal power, monitor the Oscilab visually by television, monitor the Oscilab's breathing gas and maintains communications with the Oscilab.

Submerged—Emergency

There are five situations in which all divers leave the Oscilab immediately: if the support ship must leave position quickly because of weather and sea conditions, if there is hull leakage, if there is severe contamination of the habitat's atmosphere which cannot be rectified within the BIBB system time allowance, if the life-support system fails or if there is serious illness or injury.

The following actions should be taken for the preceding situations:

1. Cut off all systems except the science refrigerator and secure the Oscilab to the extent that the situation permits. If conditions permit, switch the refrigerator to emergency power to preserve specimens until the Oscilab is reoccupied.
2. All divers are to leave the Oscilab by the wet room or emergency hatches and make a wet entry into the PTC.
3. The PTC is secured and its winch activated to pay out the cable for a controlled surfacing.
4. The PTC surfaces well clear of the support ship and is towed by a work boat to the support ship. It is taken aboard by the support ship's crane and mated to the deck decompression chamber (DDC).
5. Divers enter the DDC to complete decompression procedures.

There are three situations in which the divers do not leave the Oscilab immediately. If there is a loss of normal power or communications, the divers should switch to emergency power and reduce power load to a minimum until normal power has been restored or six hours have elapsed, whichever occurs first. If the Oscilab atmosphere is contaminated but may possibly be rectified within the BIBB system time allowance, the divers should make every effort to take corrective action if the source of trouble is aboard the Oscilab. Finally, when there is a major system failure that may be rectified within a stated time or within the BIBB system time allowance, the divers do not leave immediately.

Whenever it is ascertained that the emergency situation will exist beyond six hours or the BIBB system time allowance, whichever is applicable to the situation, preparations for surfacing in the PTC should begin immediately.

Divers. Divers follow the procedures outlined above.

Oscilab. The Oscilab carries an emergency power source of lead-acid batteries located near the keel just forward of the pressure hull. It can provide six hours of emergency power for the following:

1. Gas control equipment
2. Sonic transmitter
3. 500 watts for emergency lighting
4. Heated suits
5. Refrigeration

If items (4) and (5) above are eliminated, the emergency power would last 20 hours.

The Oscilab can be entered for emergencies or other reasons through the topside or wet-room locks. This can be done very easily by a person using shallow diving gear when the Oscilab is near the surface.

Support Ship. The support ship's principal emergency activity is to retrieve the PTC, hoist it aboard, and mate it to the DDC. The Oscilab should be immediately notified of any potential or existing factor on the surface that might contribute directly or indirectly to an emergency situation.

Submerged—Rig for Surfacing

Divers. All external equipment that is not to be left at the site is stowed aboard the Oscilab. A diver-crewman disconnects the preventer bars from the anchor and stows them near the keel. Two small electric cables, one a stand-by, are plugged into receptacles on the main anchor for remote energizing of the solenoid blow valves. The cables, carefully flaked out on the bottom, are paid out as the Oscilab rises. All hatches are closed and the divers take their surfacing stations, which are the same as their submerging stations. The record of change in load weight while on the bottom is checked and adjustments are made in the variable ballast to ensure that the winch and cable will not be overloaded. The support ship is notified that the Oscilab is ready to begin surfacing procedures.

Oscilab. The moored cable tension, C_M, should be practically the same for surfacing as it was for the submerging procedure. That is,

$$C_M = D_S - W_S$$

where

$$W_S = W + SV_{MB} \pm \text{change in load} \tag{2-5}$$

Variable ballast should be adjusted until the change in load is zero.

Support Ship. The support ship stands by to reel in the umbilical.

Surfacing—Decompression Halt

Divers. All divers are at their surfacing stations while the Oscilab is winching up from the bottom to a depth of 40 feet. During the decompression halt, the divers may move about freely and continue with their laboratory work and daily living routines. Decompression procedures are rigidly observed.

Oscilab. The pressure hull is constructed so that it may be loaded by internal or external pressure. This permits decompression procedures to take place on the surface or at any depth to 175 feet. Under normal operating conditions, decompression takes place during the decompression halt at a depth of 40 feet. Except in severe seas, relatively little motion will be caused by surface waves and useful work can therefore be accomplished during this period. The Oscilab is close enough to the surface that assistance can be readily obtained from the support ship.

The cable tension, C_D, experienced in winching up to the decompression-halt depth is expressed by the equation

$$C_D = C_M - \text{hull drag at winching-up velocity}$$

$$(2\text{-}6)$$

Hence, there is a reduction in cable tension during the ascent.

The moored tension at this level, C_M, is almost the same as it was during the surfacing phase and may be expressed by

$$C_M = D_S - W_S \qquad (2\text{-}7)$$

The submerged displacement, D_S, will most likely be somewhat smaller at 40 ft than it was at greater depths since there is a decrease in specific weight of seawater with a decrease in depth. This reduction in D_S is of no consequence since it further decreases the cable tension.

Surfacing—Interface Breakthrough

Divers. The divers secure all laboratory equipment and other loose gear and return to their surfacing stations for the interface-breakthrough phase. The diver-crew energizes the solenoid-controlled blow valves to discharge water from both sections of the anchor. If the anchor does not break free, the divers start the water-jet pump. Upon the first indication of ascent, one of the diver-crew operates the shear valve mechanism to cut the electric cables, thus permitting them to fall clear of the main cable.

Oscilab. The Oscilab ascends through the interface with an average velocity of 1.94 ft/sec.

Support Ship. The support ship prepares to tow the Oscilab and stays clear of the surfacing area.

Surface—Awash

The Oscilab breaks surface in the "awash" condition; the awash waterline is at the level of the catwalk.

Surface—In Transit

Divers. The operational cycle is completed when the diver-crew blows the main ballast tanks. The main anchor is hoisted to within a few feet of its housed position and the stream anchor is retrieved. A diver-crewman enters the topside trunk and, as a safety precaution, closes the lower hatch before opening the upper hatch. When the Oscilab has sufficient freeboard, the bottom hatch is opened. The main anchor is housed after inspection. All systems are secured and the divers prepare to leave the Oscilab.

Oscilab. When the main ballast is blown, the Oscilab approximately regains its original reserve buoyancy and floats at about the design submerging trim waterline.

Support Ship. If surface conditions permit, scuba divers from the support ship enter the water to check the main anchor for fouling and to retrieve the electric cables. Support personnel board the Oscilab for inspection. All hull penetrations are secured and the hull is again pressurized to 50 psi with air to monitor hull leakage in transit. The support ship's mooring anchors are retrieved, a towing line made fast to the Oscilab, and the tow put underway.

Principal Elements of the Oscilab Design

The principal elements considered in the design of the Oscilab are

1. Hull design
2. Main anchor system
3. Communications, power and lighting
4. Life support systems
5. Diver mobile equipment.

These have been investigated and designed in sufficient detail to ensure that the overall design is technically feasible and to enable a reliable analysis of construction and operating cost estimates.

Hull Design

The following discussion of the Oscilab hull design includes structure, materials, buoyancy, stability, a condition summary table and interior general arrangement.

Structure

This report discusses only the principal structural item, the pressure hull. The scantlings and other structural items are estimated for weight, buoyancy, and stability studies.

The pressure hull design involves two major considerations, geometry and strength. The geometry is concerned with size and shape. In the interest of low breathing-gas costs, the size is just large enough to provide for adequate working and living spaces. The shape is a function of interior general arrangement as well as of hydrodynamic considerations. A cylinder was chosen for this design because great depth and associated strength requirements are not present.

Strength requirements depend largely upon the interior and ambient sea pressure differentials and the "directions" of these differentials with respect to their manner of loading the pressure hull. A summary of pressure differentials, Δp, and their directions are as follows:

1. Depth—0 to on-site depth (300 feet is maximum); Oscilab pressurized to site-depth on surface.
   Δp—133 psi
   Direction—internal (material in tension)
2. Depth—on-site depth
   Δp—0 at interface to about 6 psi at top of hull
   Direction—internal (material in tension)
3. Depth—on-site to 40 ft (start decompression halt)

$\Delta p_{(max)}$—115psi
   Direction—internal (material in tension)
4. Depth—40 ft (end of decompression halt)
   Δp—18 psi
   Direction—external (material in compression)
5. Depth—0 ft (if decompression is on surface)
   $\Delta p_{(max)}$—133 psi
   Direction—internal (material in tension)

The greatest external Δp is 18 psi, and the greatest internal Δp is 133 psi. These two values must be compared with critical values derived from pressure-hull design data to check the structural adequacy of this data.

The severest limitation on the external Δp is imposed by a structural instability consideration. This critical value for Δp is found by using the *ASME Boiler Code for Unfired Pressure Vessels* and the following data on this design:

Plating thickness = τ = 5/8"

Stiffener spacing/hull diameter = $\dfrac{L}{D}$ = 0.555

$D/\tau = 173.0$

The code yields a critical Δp value of 75 psi which is equivalent to 169 feet. The required external Δp of 18 psi is less than 75 psi; hence, under the design conditions outlined above, the plating thickness and stiffener spacing are more than adequate. This is particularly true in view of the safety factor of 4 incorporated in the criteria of the code.

The limitation on the internal Δp value (material is in tension) is calculated by using the "Hoop-Stress" equation:

$$\sigma n = \frac{\Delta p \; r}{\tau}$$

or

$$\Delta p = \frac{\sigma n \; \tau}{r} \tag{2-8}$$

where

σn = "Hoop Stress" = $\dfrac{\sigma yp}{\text{safety factor}}$

τ = Plating thickness

r = Pressure-hull radius

The material used in this design is USS T1 steel, which has a lower yield strength value, oyp, of 90,000 psi. A safety factor of 4 was used, paralleling the *Code* procedure, in the following calculation:

$$\Delta p = \frac{90,000}{4} \times \frac{5}{8} \times \frac{1}{54} = 260 \text{ psi}$$

This is equivalent to a depth of 585 feet. The required internal Δp of 133 psi is less than 260 psi; therefore, under the design conditions outlined above, the plating thickness and the yield stress of the material used are adequate.

Material

The material used in this design, USS T1 steel, has a lower strength value of 90,000 psi and an upper-yield strength value of 100,000 psi. This steel is representative of a low-alloy, high-strength steel which has excellent corrosion resisting properties and which can be easily fabricated and welded.

Buoyancy

The buoyancy of the Oscilab hull design is considered from two points of view, surface reserve buoyancy and submerged buoyancy. The surface reserve buoyancy is the potential buoyancy of the Oscilab when on the surface and floating at the design submerging trim waterline. It is the total of all sources of buoyancy above this waterline and is equal to 27.9 long tons for this design. The reserve-to-submerged buoyancy ratio is 13.8 % which provides for about 4 ft of freeboard, measured from the submerging trim waterline to the catwalk. This height allows for safe topside working conditions and enhances the general seaworthiness of the Oscilab within the sea-state limitations imposed on the overall design concept.

The Oscilab is designed to have positive submerged buoyancy in all submerged conditions. This design relies on the tension force exerted by the main-anchor cable to maintain static equilibrium or to produce vertical motion.

The buoyancy principle of the Oscilab differs from the buoyancy principles of submarines and bathyscaphes. Submarine design almost universally employs the principle of submerged neutral buoyancy. This principle was ruled out for the Oscilab design because, in addition to sophisticated equipment required to maintain precise depth control, hydrodynamic forces and moments induced by forward motion are required to change depths while submerged. The Oscilab, of course, can only move up and down.

The bathyscaphe principle utilizes negative buoyancy for descent. The rate of descent is regulated by the controlled release of iron pellets. The total weight of the pellets is such that the total weight minus the weight of the pellets released in descent will cause the overall weight reduction required to place the bathyscaphe in nearly neutrally buoyant conditions when on the bottom. A "hovering" condition is obtained by tethering a small additional weight, a steel ball, to the bottom of the craft. Ascent is achieved by valving off an additional pellet weight. Weight may be added to the craft by valving off some of the buoyant material (aviation gasoline) and replacing it with sea water.

Certain of the preceding features might well be incorporated into a future deep-depth Oscilab design to circumvent difficulties associated with handling long lengths of cable. An untethered design incorporating horizontal propulsion devices would provide almost unlimited on-site mobility. However, it is impractical to consider the use of these concepts as long as it is necessary to connect the Oscilab and the support ship by an umbilical for two weeks. Safety and the success of the mission dictate that the two crafts be securely moored for this period.

Stability

The Oscilab's stability is considered from two points of view, surface and submerged.

Surface Stability. Transverse stability is of primary concern in surface operating conditions and is measured by

$$G_v M_t \tag{2-9}$$

where

$$G_v M_t = KB + BM_t - KG_v$$

In these expressions

$G_v M_t$ = Transverse metacentric height corrected for the "free-surface" effect of partially filled tanks

KB = Height of the center of buoyancy above the keel when Oscilab is floating at the submerging trim waterline

BM_t = Transverse metacentric radius which is equal to I/V where I is the second moment of area of the submerging trim waterplane about its longitudinal axis and V is the volume of displacement below the submerging trim waterline

KG_v = Height of the "virtual" center of gravity above the keel; this height contains a transverse free-surface correction to account for the free-surface effect of partially filled tanks

Two conflicting requirements must be considered in arriving at a value of $G_v M_t$. First, the value of $G_v M_t$ must be large enough so that, in concert with other hull parameters relating to stability, adequate static and dynamic keeling characteristics are developed. Second, the value of $G_v M_t$ must be small enough to prevent short, snappy periods of roll that result in inertia stresses on structure and equipment, as well as physical discomfort for those on board.

In the Oscilab design

$$G_v M_t = 2.95 \text{ ft}$$

in which an estimated free-surface correction (a reduction) of 0.05 ft is used. The period, T, for a complete roll is found by using the equation

$$T = \frac{2 k}{\sqrt{gG_v M_t}} = \frac{1.108k}{\sqrt{G_v M_t}} \text{ sec} \tag{2-10}$$

where k is the mass radius of gyration.

If we assume that the axis of roll passes through the center of gravity and remember that Oscilab's depth is about twice its beam, k is estimated to be 7.2 feet. The period for a complete roll is then

$$T = \frac{1.108 \times 7.2}{\sqrt{2.95}} = 4.64 \text{ sec}$$

These figures indicate that it would be desirable to reduce $G_v M_t$ to obtain a longer period of roll. However, the submerged stability requirement necessitates a low KG_v which leads to a somewhat excessive value of $G_v M_t$. Since under normal conditions no personnel are aboard the Oscilab while it is under tow, there is little need for a longer period of roll from the viewpoint of physical comfort. This short period of roll must be considered in the design of structures and equipment.

Submerged Stability. Longitudinal stability is of primary concern in submerged operating conditions and is measured by

$$BG_v$$

where

$$BG_v = KB - KG_v \tag{2-11}$$

In those expressions

BG_v = Distance between the submerged centers of buoyancy and gravity

KB = Height of the center of buoyancy above the keel when the Oscilab is totally submerged

KG_v = Height of the "virtual" center of gravity when the Oscilab is totally submerged; this height contains a longitudinal free-surface correction

In the statics of the design, the primary requirement for longitudinal stability is that it be sufficient to limit the fore and aft trim to a maximum of 1.5° when the Oscilab is subjected to trimming moments. To meet this criterion, let

$$BG_v = \frac{TM}{D_s \, Tan \, \theta} \qquad (2\text{-}12)$$

where

TM = Trimming moment, assumed to be caused by shifting a weight (the weight of six men = 175 x 6 = 1,050 lbs) a distance of 35 ft: (the "walking" length of the Oscilab pressure hull)

D_S = Submerged displacement of the Oscilab, including an average of cable tensions in all modes of submerged operation

θ = Limiting fore and aft static trim angle, 1.5°

Then

$$BG_v = \frac{1020 \times 35}{202.7 \times 2240 \times 0.0262} = 3.00 \text{ ft}$$

The actual design BG_v is 4.00 ft which allows for a longitudinal free-surface correction of 0.10 feet. Using this value,

$$Arctan = \frac{1020 \times 35}{202.7 \times 2240 \times 4.00} = 0.0196$$

or

θ = 1.1 degrees

The difference in BG_v values of 1.00 ft represents a longitudinal stability design margin to be used if submerged dynamical stability requirements so dictate.

The hydrodynamic forces and moments experienced during all modes of submerged operation and Oscilab's response dynamics are not investigated as a part of this report. A thorough study of this aspect of the hull design can only be made using data from model tests. These tests should be conducted in the next phase of the Oscilab design.

The high value of BG_v required is obtained by incorporating a large amount of buoyancy, external to the pressure hull, into the design. This permits the addition of a comparable weight of lead ballast located at the keel, thus reducing KG_v and increasing BG_v. The next phase of the Oscilab design should optimize the amount and location of this external buoyancy, giving particular attention to all aspects of enlarging the pressure hull to include some or all of the external buoyancy.

Condition Summary Table

Table 2-1 summarizes weights, displacements, and centers for the essential nonoperating and operating conditions. Data ara given in long tons (LT), ft, and foot-ton (ft-ton) units. The reference line for vertical moments and centers is the keel, and the reference line for the longitudinal moments and centers is amidships. The following notes concern this table:

1. Conditions A, B, and C are nonoperating conditions where
    a. A is Oscilab completely constructed and outfitted but having no lead, loads or variable ballast aboard.
    b. B is A plus the lead ballast and corresponds to the "light-ship" designation for ships.
    c. C is B plus the full design load (fixed and variable) aboard and sufficient variable ballast (water and solid) to compensate for changes in the variable load.
2. "Intercondition" weights and centers are given for the loads and variable ballast.

Table 2-1
Condition Summary Table
Weight and Displacement—Long Tons
Lever Arms—Feet

Condition	Description	Wt	Vert Lvr	Vert Mom	Long (F) Lvr	Long (F) Mom	Long (A) Lvr	Long (A) Mom	Disp	Vert Lvr	Vert Mom	Long (F) Lvr	Long (F) Mom	Long (A) Lvr	Long (A) Mom
A	Basic Light Condition	104.55	9.75	1019			0.60	62							
	Stability Lead	47.08	1.00	47	0.92	43	0.12	19							
B	Light Condition	151.63	7.01	1066											
	Fixed Load	2.43	8.18	20	0.70	2	0.67	12							
	Variable Load (Design)	18.07	3.45	62											
	Variable Ballast Water (Design)	1.13	6.00	7											
	Variable Solid Ballast (Design)	1.00	8.50	9											
C	Light + Load + Variable Ballast	174.25	6.66	1165	4.00	4	0.14	24							
	Cable Tension (Anchor Dry)	0.50	7.00	4											
1	Surface—In Transit (Div. Trim)	174.75	6.68	1169			0.14	24	174.75	9.33	1631				
	Change in Cable Tension and Displacement (Lower Tank Full)	+1.38	7.00	10					+1.38	15.80	22				
2	Surface—Rig for Submerging	176.13	6.69	1179			0.14	24	176.13	9.39	1653				
	Flooding Main Ballast Tanks—Change in Disposition	26.59	3.38	90					+26.52	16.82	446			0.90	24
3	Submerging—Interface Breakthrough	202.72	6.25	1269			0.12	24	202.65	10.26	2099			0.12	24
	Change in Cable Tension (Anchor on Bottom Full)	-0.07	7.00	1											
4	Submerged—Tethered (Start Winching Up or Down; On Site; Decompression Halt)	202.65	6.26	1270			0.12	24	202.65	10.36	2099			0.12	24

The cable tension (exerted by the dry main anchor in water) is added to C to obtain 1. It should be stressed that the weight of the main anchor in any condition is not considered to be a part of the Oscilab body weight. It is considered a force exerted on the body through tension in the cable. This cable is "attached" to the hull 7.00 ft above the keel at amidships.

3. Conditions 1 through 4 are the essential operating conditions that involve weight, displacement, and centers. The table also shows intercondition changes in cable tension and the weight and centers of the main ballast water. (See Figures 2-8 and 2-9.)

Oscilab Arrangement

Furnishings and arrangement for the Oscilab habitat have been designed to meet user specifications and needs and to be consistent with the engineering requirements for this undersea research facility. Figure 2-1 shows the location of all major components. The six major areas of the habitat follow.

Wetroom. The main access trunk and hatch are located at the center of the forward end of the wetroom. Adequate space for diver equipment storage, preparation and handling, and don and doff is available. A hot water shower in a closed shower stall with provision for trapping and holding all shower waste water is provided. Hookah pumps and hose-handling and storage facilities are located here, although the hookah reels should be located outside the hull adjacent to the access trunk if dependable waterproof reels could be obtained. A handrail is provided for assistance in diver entrance and exit. A live specimen tank with a circulating seawater pump is also in this area.

Laboratory. Three feet of laboratory bench space for each of the four diver-scientists and two crew members is provided. Shelves and instrument racks are above and beside this bench space. Two combination closet-lockers are available for general personal gear as well as for scientific equipment storage. The topside access trunk and lock open into this area.

Galley. Food storage and preparation areas for a two-week mission include cabinets, an electric oven and a four-burner cooktop, a freezer, a refrigerator, and a galley sink. Folding chairs and folding food trays are stored here.

Control area. All system controls including life support and lab maneuvering are located in this area.

Berthing Area. The six occupants of the lab are berthed on nylon line supported pipe berths, three high. When lowered, the middle berths serve as backs for two settees formed with the lower berths. Six storage drawers for personal gear are beneath the lower berths. The emergency trunk and hatch are located at the aft end of the berthing area.

Head and Lavatory. A separate enclosed head and an open lavatory with hot and cold water are provided. Fresh water from external rubber storage bags is supplied to a small-capacity electric water heater located in the head compartment; the heater feeds a small storage tank mounted above and somewhat behind the freezer unit.

Miscellaneous installations in the habitat include eight to ten small diameter viewing ports for general and scientific observation, general and special lighting fixtures, electrical outlets for appliances and scientific instruments, heating units, atmosphere conditioning equipment and circulating fans, communication devices, and emergency breathing gas outlets with portable second-stage regulators, hose and mouthpieces.

Main Anchor System

The principal components of the main anchor system, as shown in Figures 2-5 and 2-6, consist of the main anchor, the winch, the cable and system auxiliaries.

Main Anchor

Geometry. The main anchor, shown in Figure 2-6, has the shape of a truncated cone capped by an elliptical dome. This shape was chosen mainly because the main anchor must be drawn by a single cable into the recess at the bottom of the Oscilab. Thus, it must be circular in cross section and provided with sloping sides to ensure that it will enter the recess without fouling and to provide for snugging against suitable pads so that it will ride in a housed condition without bumping. This shape also increases the stability of the bottomed anchor.

Anchor Ballasting. The anchor is divided into upper and lower sections, each containing four ballast tanks. (See Figure 2-6.) The lower section is completely flooded while the Oscilab is on the surface. It provides the net downward force for the interface-breakthrough phase of descent. Using notations previously defined, pertinent calculations for the capacity of the lower section are as follows:

$$SV_{MB} + SV_{LA} - SV_{RB} = F$$

$$SV_{LA} = F + SV_{RB} - SV_{MB}$$

where

F = Drag force at an assumed descent velocity of 30 ft/min
= $C_D + \rho/2 \times A_{proj} \times V^2$
(C_D is established from data for cylinder and flat plates as given in Saunder's *Hydrodynamics of Ship Design*. For $R_N = 3.9 \times 10^5$ and $\frac{L}{B} = 7.1$, C_D is estimated at 1.00)
= $1.00 \times \frac{1.9905}{2} \times 650 \times (.5)^2$
= 162 lbs = 0.07 LT

SV_{RB} = Reserve buoyant force above the submerging trim waterline
= 27.90 LT, which provides for a freeboard of about 4 ft (to the catwalk)
SV_{MB} = Weight of main ballast water
= 26.59, which is approximately equal to the reserve buoyant force with the submerging trim waterline to the catwalk

Therefore,

$$SV_{LA} = 0.07 + 27.90 - 26.59 = 1.38 \text{ LT}$$

$$V_{LA} = 48.3 \text{ ft}^3$$

A very small incremental or motivating force, F, is required for the low descent speed of 30 ft/min or 0.5 ft/sec. This force can be obtained by flooding 48.3 ft³ of the 280 ft³ in the anchor ballast tanks.

The upper section of the main anchor is flooded when the anchor bottoms to provide the "hold-down" weight necessary for securely mooring the Oscilab. The following equations are used to compute the required capacity of the upper section:

$$C_M = W_A + SV_{RB} - SV_{MB}$$

$$W_F = W_A + SV_{LA} + SV_{UA}$$

$$W_F + f (C_M + L) \tag{2-13}$$

where, in addition to previously defined notations,

SV_{UA} = Weight of ballast water in the upper section
W_A = Dry weight of anchor in water
= Structure and equipment + lead ballast - displacement
= 9.00 + 0.45 - 8.95 = 0.50 LT
W_F = Fully flooded anchor weight in water
f = Hold-down allowance factor
L = Hydrodynamic lift force

Assume that the anchor has bottomed on a 20-degree slope and that a 5-knot bottom current is flowing up and under the Oscilab. Under these conditions, the vertical component of the current flowing normal to the Oscilab's axis would be 1.71 knots or 2.88 ft/sec. Then, the hydrodynamic lift force is estimated as

$$L = C_D \ x \ /2 \ x \ A_{proj} \ x \ V^2$$

$$= 1.00 \ x \ \frac{1.9905}{2} \ x \ 650 \ x \ 2.88^2$$

$$= 5360 \text{ lbs or } 2.40 \text{ LT}$$

and

$$C_M = 0.50 + 27.90 - 26.59 = 1.81 \text{ LT}$$

Therefore

$$W_F = f(1.81 + 2.40)$$

$$= f \ x \ 4.21$$

Let f be equal to 1.50 to ensure that the anchor will hold the Oscilab down under these severe conditions. Then

$$W_F = 6.32 \text{ LT}$$

and

$$SV_{UA} = W_F - W_A - SV_{LA}$$

$$= 6.32 - 0.50 - 1.38$$

$$= 4.44 \text{ LT}$$

$$= 155 \text{ ft}^3$$

It should be emphasized that the preceding calculations for the capacity of the upper section do not account for any change in the Oscilab's load while on the bottom. Load weight changes will occur of course, but these changes will be compensated for by the variable ballast system.

The following paragraphs discuss the possibility of the anchor tipping under the extreme conditions just described. Figure 2-10 illustrates this condition. The equation for the summation of moments about point B is

$$4.21 \text{ (BC)} - 2.11 \text{ (BD)} = 0 \qquad (2\text{-}14)$$

where

B = Center of gravity of fully flooded anchor in water

AB = 4.0 ft (3 ft above anchor's base)

BC = 4.0 x sin 20° = 1.37 ft

Then

$$BD = \frac{4.21 \ x \ 1.37}{2.11} = 2.74 \text{ ft}$$

and

$$DE = 2.74 \ x \cos 20^{°} = 2.57 \text{ ft}$$

DE is less than the radius of the anchor at its base, 5 ft; and PE is slightly less than DE. Therefore, the point of application of the bottom reaction, P, lies on the base area and the anchor will not tip.

An undesirable condition would exist if the prevailing currents shifted and were perpendicular to the longitudinal axis of the Oscilab. High current velocity in such a direction could tip the anchor. However, the anchor would still hold the Oscilab down even when tipped; and it could still be blown to raise the chamber.

Operation. The following are main-anchor operations, as shown in Figures 2-8 and 2-9

Surface—in transit. The dry anchor is housed in the anchor recess. It is held in place by the cable and by dogs connected by reach rods to topside hand-operated levers. A short electric cable connects the Oscilab with the lower section solenoid vent valves.

Surface—rig for submerging. The lower section of the anchor is flooded by energizing the vent-valve solenoids. The anchor is then lowered to 50 ft above the ocean bottom. The short electric cable pulls free of the Oscilab in the first few feet of the anchor's descent.

Submerging—interface breakthrough. When the main ballast tanks are flooded, the main anchor, with the lower section flooded, sinks to bottom, pulling the Oscilab through the shallow inter-

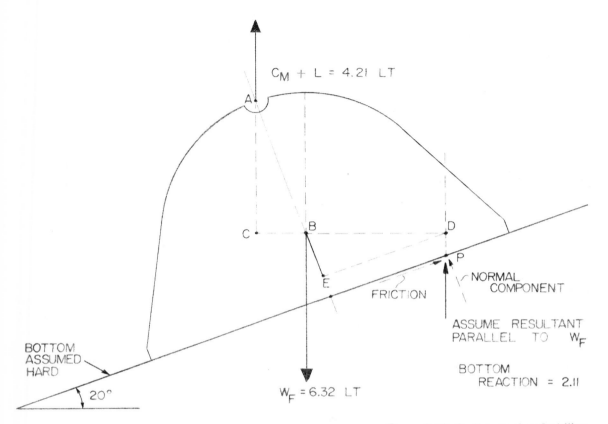

Figure 2-10. Oscilab Anchor Stability.

face depths. An electro-mechanical or pressure sensitive linkage is tripped at bottom contact, opening the upper section vents and allowing the upper section to flood.

Submerged–on site. The main anchor is fitted with pad-eyes for securing preventer bars between the anchor and the Oscilab. These bars, attached by the first diver out, preserve fore and aft trim and serve as cable back-up devices.

Submerged–rig for surfacing. The preventer bars are removed and all vents on the main anchor are closed manually. Two 400-ft light electric cables are connected to the Oscilab and the main anchor with the slack flaked out on the bottom. One cable serves as a back-up and both are capable of energizing all solenoid-operated blow valves and starting the water pump.

Surfacing–interface breakthrough. The solenoid-operated blow valves are energized, and the upper and lower section ballast tanks are blown with high-pressure air stored in air flasks in the anchor. The blown ballast issues water through nozzles designed to remove mud if the anchor has lodged in a soft bottom. The nozzles are provided with special float valves that will pass water but retain air. All possible low and high points in each tank are vented separately so that the tanks will drain and vent completely and any degree of tilt will be short of overturning.

If the anchor is not freed when the ballast is blown, a high-pressure water pump can be activated. This pump operates high-pressure water jets on the underside of the anchor to help free it. In addition, a "bubble" of air can be injected into the main ballast tanks to loosen the anchor from the bottom. Immediately after the anchor begins to ascend, these tanks should be vented so that ascent will not be too fast. As soon as the anchor is lifted clear of bottom, the two light electric cables are cut loose.

Surface. If sea conditions permit, scuba divers inspect the anchor and cable for fouling when the anchor is within a few feet of Oscilab. The electric cables are unplugged and retrieved on the Oscilab topside deck. In adverse seas, no attempt is made to salvage the electric cables. Since the anchor and cable rotate, these electric cables have little opportunity to become entwined around the main cable when falling away from the Oscilab as the anchor is lifted off the bottom. Ideally, these cables will fall clear of the anchor as it rises. Some of the cable may be looped across the anchor; however, this possibility is diminished by the bee-hive shape of the anchor. Nevertheless, it would not be a serious handicap to the anchor housing since the snubbing pads have some resiliency and can accommodate the light cable to a certain degree.

When the anchor is housed, the high-pressure line is brought topside by scuba divers where it is uncapped and connected to a source of high-pressure air. This is done to avoid getting salt water into the airlines, which might occur if an underwater connection were attempted. The anchor battery pack is replaced.

Winch

The following considerations govern the choice of the winch, the winch drive and their locations.

Lifting capacity. The greatest demand on the winch system lifting capacity occurs during the winching-down phase of submerging. This is a dynamic condition in which the cable tension, C_D, is 1.88 long tons or 4,210 pounds. The winch chosen, with a rated lifting capacity of 7,000 lbs., provides a margin of 2,800 lbs. or 40 %. The drum can hold 400 ft of 3/4-in. marine cable. The winch system (excluding the cable) weighs about 2,500 lbs; requires a space 6 ft long and 3 ft in diameter; and costs $5,000.

Powering. Two possible sources of winch power are compressed air and electricity. Compressed air drives are very inefficient and therefore demand large quantities of compressed air, which is a luxury in this design since all high-pressure air is carried aboard the Oscilab. On the other hand, electric power is supplied from the support ship under normal conditions, and electric motor drives are more efficient than air drives. A 10-horsepower electric motor which operates under ambient pressures in an oil bath was therefore selected.

Location. The winch and winch drive are located topside just aft of the topside trunk to facilitate maintenance and cable inspection while the Oscilab is on the surface.

Cable

The maximum "static" loading on the cable during the submerging-surfacing operations is 1.88 long tons or 4,210 pounds. If a factor of 10 is used to account for static overloads and dynamic cable loadings, the strength requirement, C, becomes

$$C = 10 \times 4210 = 42,100 \text{ lbs}$$

Specifications for the cable are

1. Diameter—3/4 in
2. Breaking load—57,800 lbs
3. Elastic limit load—43,000 lbs
4. Weight/ft—0.82
5. Type—3 x 19, Monitor AA, independent wire-rope center, torque-balanced

System Auxiliaries

In addition to controls and basic instrumentation, the principal system auxiliaries are the tensiometer and the hydraulic cable cutter. The tensiometer provides data on cable tensions. This information is necessary for the safe and proper operation of the Oscilab. The cable cutter is operated by a hand pump as a means of cutting the cable if the winch system fails.

The calculations contained in this section are vital for safe operation of the Oscilab and are

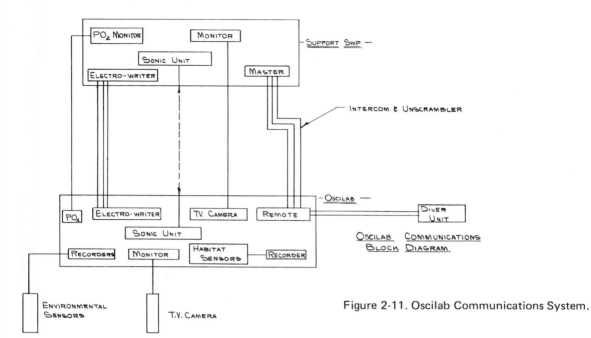

Figure 2-11. Oscilab Communications System.

Communications

based on assumed values for drag coefficients and dynamic cable loading factors. In the refinement of this design, it is recommended that these essential data be derived from extensive model tests.

Communications

The Oscilab communications system consists of links between the Oscilab and the support ship and between the Oscilab and the diver. There are audio and visual links between the Oscilab and the support ship, including interior and exterior environment sensing and recorder displays, and a back-up acoustic communication unit. An audio communications link is provided between the Oscilab and the diver; the diver's hookah gear umbilical contains the cable.

Communication and instrumentation equipment required are listed below. (See Figure 2-11 for a diagram of the communication system.)

1. Oscilab
   a. Intercom and speech unscrambler—two-way intercom with an integrated speech unscrambler linked

with the support vessel and the diver's communication unit

   b. Electrowriter—means of transmitting diagrams and figures
   c. TV camera—camera for observing the Oscilab interior from the support ship
   d. TV camera and monitor—equipment for observing and monitoring the site environment
   e. Oscilab interior environment instrumentation—chart-recording equipment for wet and dry bulb temperatures
   f. Acoustic communications unit and transducer—equipment for emergency communications between the Oscilab and support ship
   g. Environment sensor system—equipment that provides continuous chart-recording of site environmental characteristics such as temperature, pressure, current speed and direction, ambient light, pH and turbidity

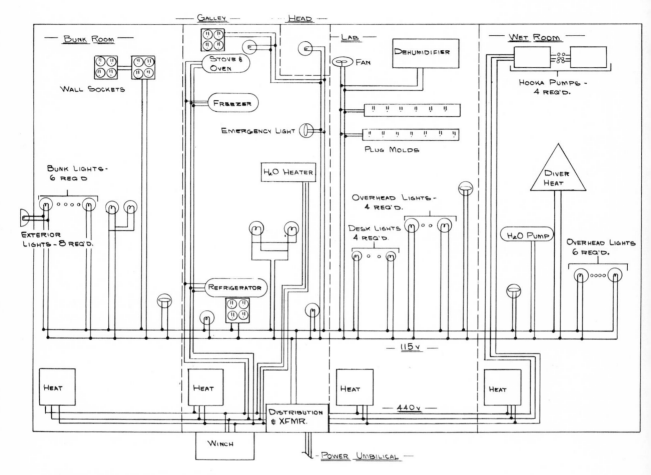

Figure 2-12. Oscilab Electrical System.

2. Support ship
    a. Intercom
    b. Electrowriter
    c. Oscilab interior TV monitor
    d. Acoustic communications unit
    e. pO_2 Monitor-equipment to monitor O_2 from Krasberg O_2 control unit

Power and Lighting

The Oscilab system has two sources of electrical power, normal and emergency. The normal source of power is a 60-kw., 3-phase, 450-volt generator located on the deck of the support ship. The power is delivered to the Oscilab through cables in the umbilical. A second generator serves as a standby unit. The emergency source of power is carried on the Oscilab and consists of a 150-kw-hr lead-acid battery pack.

The electrical system is diagrammed in Figure 2-12. Its power requirements can be divided into the following categories:

Lighting	9.6 kw
Communications	0.5 kw

Heat (interior)	13.0 kw
Heat (hot water)	4.5 kw
Refrigerators-freezers	0.2 kw
Fans	0.1 kw
Pumps	0.4 kw
Winches	10.0 kw
Diver	
Hookah pumps	3.2 kw
Heat	1.5 kw
Dehumidifier	4.0 kw
Cooking	9.5 kw
TOTAL	56.5 kw

Internal lighting is provided by 25, 50 and 75-watt bulbs (heavy duty) in totally enclosed but vented fixtures. Exterior lighting is furnished by eight 1000-watt lights placed uniformly around the Oscilab.

Life Support Systems

Atmosphere

The Oscilab interior atmosphere is a mixture of helium and oxygen. The oxygen is introduced from external bottles and mixed with helium in quantities regulated by a Krasberg pO_2 meter that maintains the desired oxygen concentration. The interior is pressurized to site depth while Oscilab is on the surface; the helium is supplied by the support ship. When the Oscilab is on site and the lower trunks are open to the sea, additional helium is introduced from external bottles to compensate for gas losses and to maintain a constant water level in the trunks. A float control in the trunk regulates the introduction of this makeup helium. The atmosphere is circulated through a purification system to remove water, carbon dioxide, hydrocarbons and other trace contaminants.

In the design of the mixed-gas system, the gas-loss-versus-reclamation ratio during the decompression process was considered. The cost of the gas required to pressurize the chamber is under $1,160. The cost of a gas recovery and purification system, estimated to be about $40,000, precluded this alternative. Hence, the gas will be discharged

directly into the air if decompression is carried out at the surface and into the sea if this procedure is accomplished at depths of 40 to 50 feet. In the latter case, a pump or surface-buoyed gas line must be used during the final stages of decompression.

The divers use a closed hookah system which delivers breathing gas from the Oscilab's interior and returns it for purification. In addition, each diver carries emergency bottles of breathing gas. A BIBB system is installed for emergencies occurring within the Oscilab.

The helium and oxygen bottles are carried in the Oscilab's superstructure, as shown in Figure 2-4. Figures 2-13 and 2-14 are diagrams of the mixed-gas system.

Moisture Control

Three ways of controling the moisture content of the sealab interior were considered: electric-mechanical dehumidifiers such as were used on Sealab II, use of the hull's surface as a condenser, and an absorption system using silica gel.

The electric-mechanical dehumidifiers were not overly successful in Sealab II. The high thermal conductivity and heat capacity of the helium-oxygen atmosphere seem to reduce the capacity of these units below the capacity obtainable in air at 1 atmosphere of pressure.

In the second system, water was removed by condensation from a section of uninsulated hull or by a condenser fed with seawater. This system was undesirable because of high power levels and cost of reheating the atmosphere. In addition, the requirement that Oscilab be capable of operating in seawater temperatures of 30° F to 85° F eliminated this alternative.

It was decided to use an absorption system for water removal. This method is simple, effective, and low in capital and operating costs. Moisture has been removed from gases in industrial processes by absorption on activated alumins and silica gel and, more recently, by molecular sieves. This absorption method has also been proved effective for preventing corrosion on "moth-balled" ships. Silica gel was selected for the Oscilab system be-

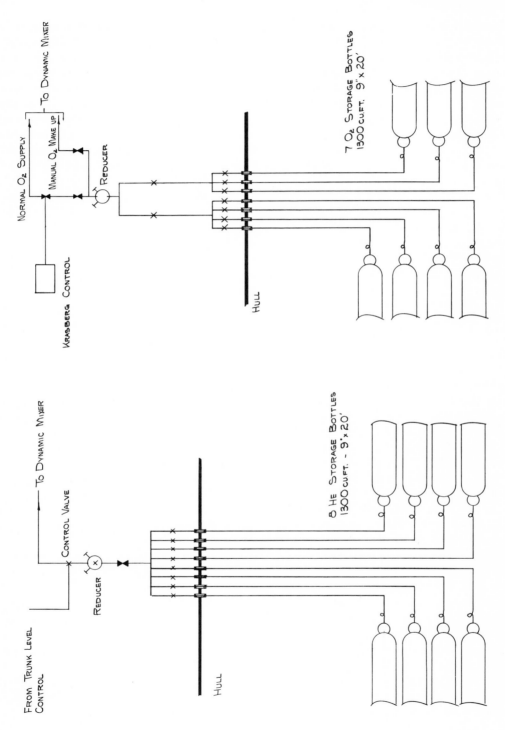

Figure 2-14. Oscilab Oxygen System.

Figure 2-13. Oscilab Helium System.

cause it has a high absorptive capacity under Sealab conditions.

The moisture-removal system, shown in Figure
2-15, consists of two chambers filled with silica gel
impregnated with a material to indicate when the
gel has absorbed the maximum amount of moisture. One chamber removes moisture from the atmosphere gas passing through it while the contents
of the other chamber is being regenerated. Regeneration consists of heating the gel to about 450°F

to distill the absorbed water. This water is con-
densed either on a few square feet of uninsulated
hull surface or in a small seawater condenser. The
gel is heated by electric heating elements imbedded in the material. The absorption system could
be operated automatically, but system-simplicity is
improved by manually shifting the chambers from
"absorption" to "regeneration" as determined by
the color of the gel. The system is designed to
operate under the most severe conditions expected

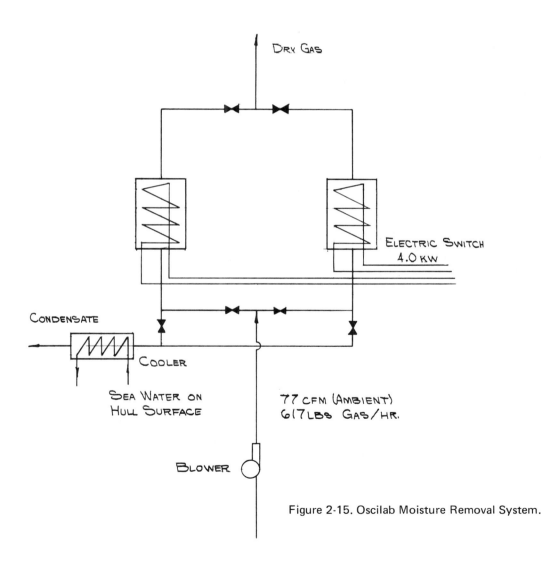

Figure 2-15. Oscilab Moisture Removal System.

for one hour before regeneration is necessary. Under normal conditions, regeneration every two or three hours is sufficient.

Carbon Dioxide Removal

Two solid CO_2 absorbents, lithium hydroxide and baralyme, were considered for removing carbon dioxide. Lithium hydroxide has the disadvantages of being very expensive, requiring a large volume for a 14-day mission, being difficult to handle if wet, and being an irritant to mucous membranes in the dry condition. On the other hand, baralyme costs much less but has a lower CO_2 absorbing capacity and a higher volume requirement for the 14-day period.

Further investigation indicated that a simple, inexpensive CO_2 absorption system using a sodium hydroxide solution could be used. Since CO_2 is an acid, its solubility is very high in alkalis because it is converted into sodium carbonates. Seawater is

slightly alkaline (pH 7.2 to 7.4) and its pH can be greatly increased by the addition of sodium hydroxide. For the Oscilab design, about 200 gal of 50% sodium hydroxide would be ample for a 14-day mission. This solution would be kept in a rubber bag outside the hull. It would be mixed with about 50 lbs of seawater per hour to increase the volume of the liquid phase. This mixture would be brought into contact with the atmos-

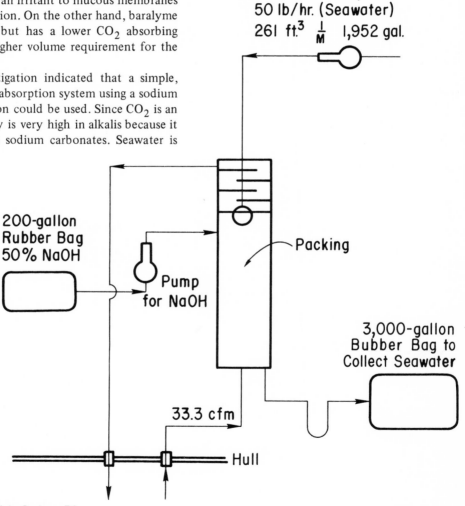

Figure 2-16. Oscilab Carbon Dioxide Removal System.

phere, which would be circulated through the mixture at 33 ft^3/min in a counter-current absorption tower. If necessary, the effluent could be collected in a 2,000-gal rubber bag outside the hull to prevent contamination of the site. The system, shown in Figure 2-16, would be continuous in operation and would require no attention during the mission.

The CO_2 content of the atmosphere is monitored continuously, using a Beckman CO_2 analyzer or equivalent equipment.

Heating and Ventilation

The Oscilab interior heating system is designed so that there are no cold surfaces to absorb radiant energy from the occupants. Heat is s provided by imbedded cables in the floor and radiant panels on the sides and overhead. Three kilowatts are provided for floor heating, 7 for side heating, and 3 for overhead heating. The floor elements would be thermostatically controlled as would be the individual radiant panels of the side and overhead system. (See Figure 2-17 for a diagram of the overall system.)

The Oscilab interior ventilation system, shown in Figure 2-18, consists of ducts with intakes near the floor and several outlets near the overhead. The occupants can adjust these outlets to vary the quantity and direction of the mix-gas flow. During each recycle, a portion of the gas will be passed through the water and carbon dioxide removers and the Purafil K (or activated carbon) filters to remove odors and trace irritants.

Sanitation and Waste Disposal

The Oscilab sanitation system, shown in Figure 2-19, includes a stall shower, a small lavatory, and a toilet. The shower provides the principal means for the divers to get warm upon returning to the Oscilab from on-site excursions. Consequently, large quantities of hot water are required; this water is carried in a 50-gal tank with a 4.5-kw heater.

The design requirement regarding strict limits on site contamination necessitates providing for the collection of all wastes and effluents. A 6,000-gal rubber bag and an 1,800-gal bag are installed outside the main hull for this purpose.

Water Supply

Two separate supplies of fresh water are provided: potable water for drinking and working and water for showers and other cleaning purposes. A stainless steel tank inside the hull contains about 170 gal of potable water. An externally located 6,000-gal rubber bag contains water for showers and other cleaning purposes. These capacities should be sufficient for six men on a 14-day mission.

Food Preparation and Preservation

Ample food, including dehydrated and frozen food, is provided for a 14-day mission. Principal cooking equipment includes an electric stove and radiant heat or micro-wave oven. These facilities are vented so that the cooking odors will not contaminate the atmosphere. A combination freezer-refrigerator provides for food preservation.

PTC Life Support System

The PTC has a completely independent life-support system capable of supporting six men for 6 hours under emergency conditions. Breathing gas is contained in two size A helium bottles and one size A oxygen bottle. A blower and duct system will circulate the gas through baralyme and Purafil K for carbon dioxide and hydrocarbon removal. No provision is made for moisture removed or heating. Emergency kits of water, food and medical supplies are stowed in the PTC.

Diver Mobile Equipment

Breathing Apparatus

The diver's breathing gas may be derived from two sources while he is on site excursions away from the Oscilab, the normal source and the emer-

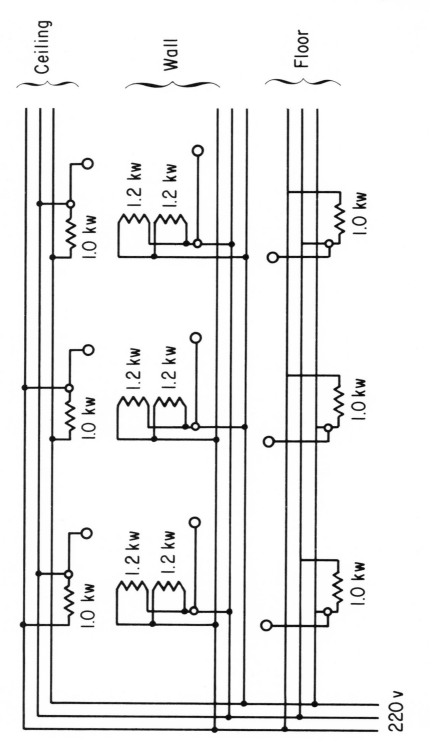

Figure 2-17. Oscilab Heating System.

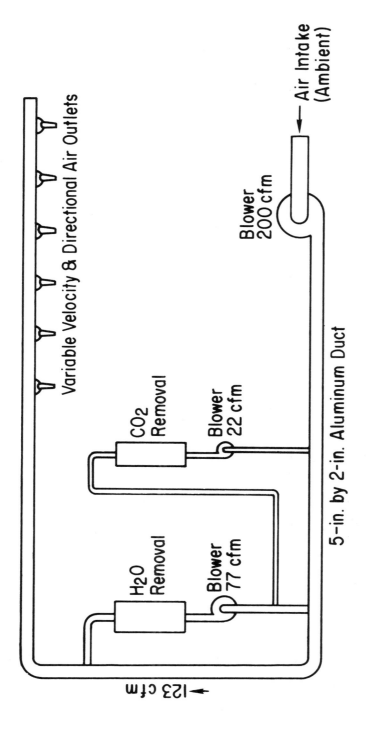

Figure 2-18. Oscilab Ventilating System.

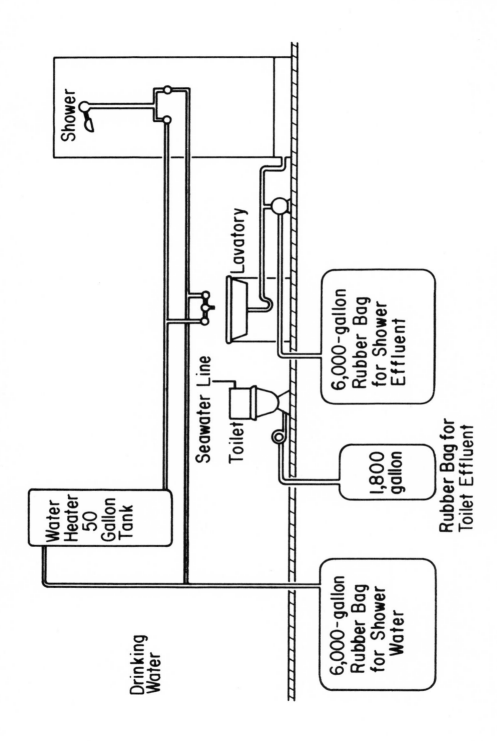

Figure 2-19. Oscilab Water and Sanitation System.

gency source. The normal source of supply uses a closed-circuit hookah system in which the gas is pumped from the interior through one line of a double hose to the diver. After inhalation, the diver exhales into the other line of the hose, and the gas returns to the Oscilab for purification and revitalization. In addition to the double hose, the hookah umbilical contains a diver-suit heating source (a power cable or warm-water supply tube) and a communications line. The umbilical has neutral or a slightly negative buoyancy and is coiled on reels mounted on the Oscilab.

The emergency source of breathing gas is carried by the diver; it is a small-capacity, semiclosed-circuit scuba system. Two standard 6-inch-diameter, 19-inch-long D-1 bottles are charged with 22 % oxygen to 2,000 psi. A flow setting of 35 lpm allows for a gas supply duration of about 45 minutes at 300 feet. This self-contained system anticipates the possibility of the following emergencies:

1. Failure of hookah pumps due to malfunction or power loss
2. Fouling of hookah umbilical due to kinking, snagging or collapse
3. Cutting of the umbilical
4. Failure of other components of hookah system
5. Contamination of Oscilab's atmosphere

In an emergency, a quick-disconnect device in the hookah system permits rapid transfer to the scuba system.

A full-face mask with communication microphone and receiver and designed for both the closed-circuit hookah and the semiclosed-circuit scuba systems is required.

Thermal Protective Suits

Two types of diver suits, the electrically heated and the warm-water heated suits were considered. Despite the relatively high cost of electric suits, it did not seem feasible at this time to use warm-water suits because this system requires 30 to 50 kws of energy for two suits. Electric suits require 1 to 2 kw of energy for two suits. A diver-mounted,

electrically-powered, compact heat exchanger and pump could make the warm-water heated suit more attractive, but such a system is not practically available. Consequently, an electric resistance-wire heated suit with power supplied from Oscilab at 24 volts (AC) is used.

Another advantage of the electric suit is that this system eliminates the discharge of heated water into the site environment. In addition, an umbilical containing an electric cable is less bulky than one containing a warm-water supply tube or hose.

Other Equipment

Additional diver equipment including fins, watches, compasses, and emergency sonic homing devices are similar to conventional types. The equipment was designed to operate in a high-pressure helium atmosphere.

Oscilab Design Capital and Operating Costs

One of the objectives of the University Sealab study was to develop a conceptual design in sufficient detail to estimate capital and operating costs. This cost study was made under the Office of Naval Research Contract NONR 3710-04 and was published as *EDAL Technical Report No. 101*.

The costs were predicted on the following bases:

1. Labor and material costs were based on September 1966 data.
2. The surface support ship (the Sea Truck) was to be leased.
3. Operating days per year = 228
4. Average length of mission
   a. days on site = 14
   b. days in transit and set-up time = 5
   c. average total length of mission = 19
5. Missions per year = 12
6. Diving hours, scientist-diver/day = 5 to 6
7. Diving hours, scientist-diver/mission = 75
8. Research diving hours/mission = 300 man-hours
9. Research diving hours/year = 3,600 man-hours
10. Research diving days/year = 168

The following is a summary of capital and operating cost estimates for the Oscilab system:

Capital Investment

Oscilab	$732,000
Surface support systems (on ship)	58,000
TOTAL	$790,000

Operating Costs	Annual	Cost/Day
Supplies and expendables	$103,000	$ 450
Personnel and administrative	188,000	825
Maintenance	44,000	195
Leased services	420,000	1,840
TOTAL	$755,000	$3,310

Cost for each operating day was based on 228 operating days in a year. Leasing arrangements of support ship precludes nonoperating days.

Acknowledgments

The University Sealab study, the basis for this paper, involved all of the human resources of the University of New Hampshire Engineering Design and Analysis Laboratory (EDAL). Faculty and students from several engineering disciplines worked together to find a solution to this fascinating design problem in ocean engineering. The author gratefully acknowledges their individual and collective contributions and wishes to define his own area of interest—the naval architectural aspects of this overall engineering design effort.

The Office of Naval Research provided the opportunity for involvement in the scientific sealab aspect of its Man-in-the-Sea program. The author appreciates the guidance and support received from this source as well as contributions of the Scientific Advisory Committee, the Diver Consultant Group and individuals in government and private industry associated with saturation-diving facilities.

The author appreciates the help of the Center for Engineering Research of the University of Hawaii in facilitating the publication of this paper.

part 3: salvatore comitini

Economic, Institutional, Legal Aspects of Marine Resource Development

At the University of Hawaii, Dr. Salvatore Comitini is an associate professor in the Department of Economics and staff member in the Economic Research Center and Hawaii Institute of Marine Biology.

From 1966-67 he served as the Advanced Mershon Research Fellow at Ohio State University and in 1960 as the Ford Foundation Research Fellow in Economic Development at the University of Washington. Other honors include membership in Who's Who in the West, American Men of Science and Pi Gamma Mu (national social science honorary). Dr. Comitini also belongs to American Economic Association, Econometric Society, Marine Technology Society and Western Regional Science Association.

He has served as consultant to the Institute of Social, Economic and Government Research at the University of Alaska; the Division of Economic Research, Bureau of Commercial Fisheries, U.S. Department of the Interior; Fisheries Division, Organization for Economic Cooperation and Development, Paris, France; and Fisheries Department, Food and Agriculture Organization of the U.N., Rome, Italy.

Besides publishing articles, Dr. Comitini has participated in numerous conferences and presented papers. He obtained his B.S. (1951) and M.S. (1955) from the University of Alabama and his Ph.D. (1960) from the University of Washington.

3

economic, institutional, legal aspects of marine resources development

Marine Science and Resources Policy

The United States' stake in the uses of the sea is clearly pronounced in the opening paragraph of the report of the Presidential Commission on Marine Science, Engineering and Resources:

> How fully and wisely the United States uses the sea in the decades ahead will affect profoundly its security, its economy, its ability to meet increasing demands for food and raw materials, its position and influence in the world community, and the quality of the environment in which its people live.[1]

Three important factors have stimulated this recent interest in improving the nation's technological effectiveness in the uses of the sea: (1) the potential undersea threat to our national security; (2) a recognition of the significance of science and technology as instruments of international affairs; and (3) an increased international awareness and interest in the potential benefits from exploiting marine resources.

The year 1966 marks a significant turning point in our national policy regarding exploration and development of marine resources. In that year, Congress passed two important acts: "The Marine Resources and Engineering Development Act" and the "National Sea Grant College and Program Act." The former created a National Council on Marine Resources and Engineering Development and an advisory Commission on Marine Science, Engineering and Resources—both charged with developing a comprehensive and coordinated national program to advance the marine sciences. The latter act was passed as an amendment to the former when it soon was recognized that attainment of many of the objectives of this "new thrust seaward" critically depended upon the availability of a group of skilled people to apply marine science and engineering techniques, equipment and research to developing the sea's resources.

The Marine Resources and Engineering Development Act of 1966 declared [it] "to be the policy of the United States to develop, encourage, and maintain a coordinated, comprehensive, and long-range national program in marine science for the benefit of mankind to assist in protection of health and property, enhancement of commerce, transportation, and national security, rehabilitation of our commercial fisheries, and increased utilization of these and other resources." Related to achievement of these national goals, this mandate identifies the following eight objectives:

1. The accelerated development of the resources of the marine environment.
2. The expansion of human knowledge of the marine environment.
3. The encouragement of private investment enterprise in exploration, technological development, marine commerce, and economic utilization of the resources of the marine environment.
4. The preservation of the role of the United States as a leader in marine science and resource development.
5. The advancement of education and training in marine science.
6. The development and improvement of the capabilities, performance, use, and efficiency of vehicles, equipment, and instruments for use in exploration, research, surveys, the recovery of resources, and the transmission of energy in the marine environment.
7. The effective utilization of the scientific and engineering resources of the Nation, with close cooperation among all interested agencies, public and private, in order to avoid unnecessary duplication of effort, facilities, and equipment, or waste.
8. The cooperation by the United States with other nations and groups of nations and international organizations in marine science activities when such cooperation is in the national interest.[2]

The National Sea Grant College and Program Act declared "that in view of the importance of achieving the earliest possible institution of signifi-

cant national activities related to the development of marine resources, it is the purpose of this title to provide for the establishment of a program of sea grant colleges and education, training, and research in the field of marine science, engineering, and related disciplines."[3]

Thus, the act has three purposes: (1) to increase and strengthen the trained manpower pool for participation in marine resources development; (2) to strengthen and improve applied research in the marine sciences, thereby increasing the knowledge and skill required to develop new techniques and equipment for effectively developing the marine resources; and (3) to strengthen and improve the process of information transfer (e.g., marine advisory programs) in assisting persons or enterprises to further expand technology in marine resources development.

National Policies and Objectives

The Marine Resources Act requires that the President shall transmit an annual report to the Congress "which shall include (1) a comprehensive description of the activities and the accomplishments of all agencies and departments of the United States in the field of marine science during the preceding fiscal year, and (2) an evaluation of such activities and accomplishments in terms of the objectives set forth pursuant to this Act." The national policy was officially enunciated in the President's first report to the Congress:

> The resources of the oceans can help us meet many of the challenges that face our Nation and the world today.
> The vast food reserves of the sea must be developed to help end the tragic cycle of famine and despair.
> The continuing pollution and erosion of our seashores, bays, estuaries, and Great Lakes must be arrested and reversed to safeguard the health of our people and to protect the resources of the sea.
> The influence of oceans on the environment must be understood so that we may improve the long-term forecasting of storms, weather, and sea conditions; protect life and

TABLE 3-1
Total Federal Marine Science Program by Major Purpose[a]
(In millions of dollars)

	Estimated FY 1968	Estimated FY 1969	President's Budget FY 1970
International Cooperation and Collaboration	9.6	9.5	11.7
National Security	119.9	128.1	143.0
Fishery Development and Seafood Technology	40.1	43.8	44.5
Transportation	11.1	10.6	18.6
Development of the Coastal Zone	27.6	29.1	29.6
Health	5.3	5.4	5.3
Nonliving Resources	7.3	8.0	8.6
Oceanographic Research[b]	78.1	93.3[c]	93.5
Education	7.0	7.5	9.2
Environmental Observation and Prediction	28.8	31.6	32.1
Ocean Exploration, Mapping, Charting and Geodesy	75.7	83.0	99.1
General Purpose Ocean Engineering	19.2	18.3	29.5
National Data Centers	2.1	2.3	3.2
Total	431.8	471.5	528.0

[a] Many programs of the Departments of Defense, Commerce, Interior and Transportation and other agencies related to marine science affairs are not included.

[b] Research beneficial to more than one of the other major purpose areas.

[c] Includes $14.5 million for Coast Guard sub-polar oceanographic research ship. Funding for other ships is included in the relevant major purpose category.

Source: The third report of the President to the Congress on Marine Resources and Engineering Development, *Marine Science Affairs—A Year of Broadened Participation*, 11-12.

property in coastal areas; and improve the prediction of rainfall in the interior.

The wealth of the ocean floor must be freed for the benefit of all people.

Finally, the seas must be used as pathways to improved international understanding and cooperation.[4]

New Initiatives and Areas of Increased Emphasis

The first report of the council largely emphasized the transition from basic ocean science to application of science and technology towards solving pressing problems and enhancing the practical benefits to be derived from ocean use. It also accented a transition of ocean sciences considerations largely at the program level to a new orientation and responsibility at the governmental policy level.

The second report related the potential of ocean exploration and development to major national goals and presented a broad range of policy proposals and action programs to aid in their achievement. After extensively evaluating ongoing marine science activities and future needs, the council designated the following initiatives for special emphasis and priority:

1. Accelerate the nation's efforts to *expand international cooperation in ocean exploration*—thus contributing to international cooperative efforts in exploration and development of the living and nonliving resources of the sea for the benefit of mankind.

2. Accelerate the development of *food-from-the-sea* in fighting the War on Hunger—especially in developing the feasibility of producing a low-cost fish protein concentrate.

3. Foster *rational development of the coastal zone*—by emphasizing Federal-Regional-State-local cooperation in research and development and systems analysis of problems related to pollution, economic development, and conflicting uses of the coastal zone.

4. Prepare and plan new programs for *port development and redevelopment*—including comprehensive planning for incorporation of new technologies in a national port system to serve ocean shipping of the future.

5. Institute new measures for insuring *safety of life and property* off our coasts in view of the intensifying coastal traffic—including oil pollution contingency plans, designation of sea lanes and improved ship navigation systems, and establishment of safety standards for offshore structures and safe procedures and rescue services for submersible and underwater activities.

6. Increase *sea grant program* investments—for training the specialized manpower and developing the marine technology required for the future needs of ocean development.

Other initiatives deserving increased emphasis were identified as: marine application of new technology (e.g., spacecraft and buoy technology) of benefit to science and industry; deep ocean technology essentially to meet future military requirements and development of deep sea resources; arctic and subpolar research programs to sustain a U.S. capability of potential strategic and economic significance in these regions; and mapping of the continental shelf to assist in identifying the resource potential of oil, gas and mineral deposits and thus help to stimulate private investment in development.[5]

The third report focused primary attention on policy issues significantly affecting economic growth, rational development of the coastal environment, national security matters and international affairs. (See Tables 3-1, 2 and 3.) In the words of President Lyndon B. Johnson to the Congress: "While modern science and technology afford powerful means to translate marine potentialities into realities, the economic, social, legal and political considerations are equally decisive in utilizing the seas to achieve the goals and aspirations of our society."

TABLE 3-2
Total Federal Marine Science Program
by Department and Independent Agency[a]/
(In millions of dollars)

	Estimated FY 1968	Estimated FY 1969	President's Budget FY 1970
Department of Defense	240.6	262.3	297.9
Department of the Interior	70.5	76.2	78.2
National Science Foundation	38.1	35.0	45.4
Department of Commerce	33.6	33.4	49.0
Department of Transportation	15.4	34.1	32.3
Atomic Energy Commission	13.8	10.4	11.4
Department of Health, Education and Welfare	6.5	6.9	6.8
Department of State	6.6	7.0	7.1
Agency for International Development	3.0	2.5	4.6
Smithsonian Institution[b]/	1.9	1.9	2.0
National Aeronautics and Space Administration	1.8	1.8	1.7
Total	431.8	471.5	528.0

[a]/ Many programs of the Departments of Defense, Commerce, Interior and Transportation and other agencies related to marine science affairs are not included.

[b]/ Excess Foreign Currency Funds are not included.

Source: *Marine Science Affairs—A Year of Broadened Participation,* 11-12.

To assist in meeting the goals set forth in the Marine Resources Act, the council considered the following policy issues:

1. Planning for the launching of the *International Decade of Ocean Exploration* proposed by the President.

2. Development of a *legal regime for the deep ocean floor* which would encourage development of seabed resources and minimize international conflict.

3. Encouraging the states to give greater consideration to *Coastal Zone planning and management* including multiple-use activities.

4. Possible steps to strengthen the *harvesting sector of the domestic seafood industry* to lower costs of operations and increase efficiency in marketing and distribution.

5. Integrating the *development of fish protein concentrate* technology as a part of the Food-from-the-Sea Program to assist the developing nations in combatting protein malnutrition.

6. Encouraging *development of fuels and minerals* on the continental shelf and slope by private industry, including an examination of offshore leasing policies, harmful environmental side effects, and reconnaissance surveys.

7. Recognizing that successful research and development is critically related to an improved *capability for management of marine science information* through use of automated equipment and study of data needs.

Other issues deserving increased and continuing emphasis were: improving the international organizational structure for planning and carrying out marine science programs of interest to the international community; intensifying deep ocean technology research and development; instituting new measures to insure safety of life and property along our coasts and on the high seas; preparing for development and redevelopment of U.S. ports and harbors; formulating policies and programs for the arctic region; emphasizing sea grant investments for training specialized manpower; fostering marine applications of new technology; and strengthening the nation's base of marine research and technology.[6]

The federal marine science budgets for fiscal years 1968, 1969 and 1970 are shown in the following tables. These are broken down by major purpose, by department and independent agency and by function.

Food from the Sea: Problems and Potential

Food experts keep reminding people that the rate of increase of the food supply is barely keeping pace with world population growth—moreover, that one-half of the world's population—1.5 billion people—is afflicted by protein malnutrition. In view of these unsettling statistics, food from the sea offers the high promise of meeting a part of this worldwide need.

The current U.S. policy with regard to marine resources is aimed at both rehabilitating the fishing industry into a viable and efficient sector of the economy and accelerating the contribution of marine resources to the continued expansion of the economy. The policy is also aimed at encouraging international cooperation in exploiting the full potential of the seas through application of modern science and technology and economic assistance. To this end, the United States is embarked on a long-range program to exploit the food potential of the oceans to help feed the undernourished people of the world. This is in accord with the broad strategy of the War on Hunger Office of the Agency for International Development. A food-from-the-sea service was organized within the agency to assist in the technological development and distribution of FPC (fish protein concentrate)—a tasteless, odorless protein concentrate which can be combined with flour, cereals and

TABLE 3-3
Total Federal Marine Science Program by Function
Summary
(In millions of dollars)

	Estimated FY 1968	Estimated FY 1969	President's Budget FY 1970
Research and Development	245.4	248.3	297.9
Research (basic and applied)	116.4	126.7	150.6
Development of new equipment and technology	129.0	121.6	147.3
Investment	55.5	77.8	70.1
Ships	7.6	15.1	2.3
Major equipment	29.3	42.4	47.5
Shore facilities	13.9	15.5	13.3
Other	4.7	4.8	7.0
Operations	130.9	145.4	160.0
Surveys	106.0	118.7	130.9
Services	18.4	19.5	21.4
Other	6.5	7.2	7.8
Total	431.8	471.5	528.0

Source: *Marine Science Affairs—A Year of Broadened Participation*, 11-12.

other processed foods. Pilot production operations have shown the feasibility of producing ten grams of FPC—supplying the minimum daily animal protein nutritional requirement—at less than one cent.

Estimates for increasing the world fisheries catch with current techniques range anywhere from a doubling of the present output of approximately 60 million metric tons to possibly five or more times.[7] Although there are vast amounts of potentially exploitable resources off the coasts of North America, relatively few species are actually caught by U.S. fishermen.[8] Since 1956 the United States has dropped from second to sixth place among the world's fishing nations. (See Table 3-4.) What are some of the factors cited as reasons for this relative decline? Basically, they are subsumed under what may be termed "institutional constraints." That is, laws, customs, and economic regulations—national and international—which limit the fish-catching ability and capacity of fisher-

TABLE 3-4
World Catch ᵃ/of Fish, Shellfish, Etc. by Leading Countries
1954-66
(In billions of pounds)

Country Ranking by 1966 Catch Figures	1954	1956	1958	1960	1962	1964	1966ᵇ/
Peru	0.4	0.7	2.2	7.9	15.3	20.2	19.4
Japan	10.0	10.5	12.1	13.7	15.1	14.0	15.6
China	5.1	5.8	8.9	12.8	NA	NA	NA
USSR	5.0	5.8	5.8	6.7	8.0	9.9	11.8
Norway	4.5	4.8	3.2	3.4	2.9	3.5	6.3
United States	6.1	6.5	5.9	6.2	6.6	5.8	5.5
World	60.5	67.1	72.4	87.1	102.0	114.0	125.0

ᵃ/ Total commercial catches, including fresh water fish, on a live weight basis. Weights of shells of univalve and bivalve mollusks are included.

ᵇ/ Preliminary.

NA = Not available.

Source: The second report of the President to the Congress on Marine Resources and Engineering Development, *Marine Science Affairs—A Year of Plans and Progress,* 222.

men. These might include restrictions on fishing area, season, or types of vessels, gear and techniques allowed.

What rationale is there for imposition of these impediments on the efficiency and effectiveness of fishermen? Some of the reasons given are: (1) to maintain a sustainable fish catch; (2) to reconcile conflicting user interests, e.g., between sport and commercial fishermen; (3) to insure use of American-built fishing vessels; or (4) to insure that small, individual fishermen are not placed at a competitive disadvantage compared with larger fishing organizations. These policies have resulted in serious economic inefficiencies imposed on American fishing activities. They have led to the fragmentation of the industry into small units, finding it difficult to accumulate capital in sufficient amounts to adopt the latest equipment and techniques. The only real distant-water operations of the U.S. fishing fleet are the tuna and shrimp fisheries.

This approach to the U.S. fisheries policy has conditioned the strategy of the fishing industry to continually seek governmental assistance through subsidies and protection from foreign competition, rather than adoption of more efficient fishing techniques and locating new resources. Federal government assistance to the fishing industry tends to reflect this kind of strategy.

A breakdown of the U.S. Bureau of Commercial Fisheries budget for programs to foster the fisheries' growth and diversification shows the following allocation of expenditures: 63% devoted primarily for gathering data on abundance and distribution of fish stocks and attempting to estimate maximum sustainable yields; 20% applied to research into processing and marketing; while only 10% is directed to development of new fishing technology for greater catching efficiency or locating techniques, or new strategies of fishing.[9] It is generally felt by critics of U.S. fisheries policy that these constraints, resulting in limited economic viability of the fishing industry, have hampered the effective adoption and application of recent advances in ocean science and technology.[10]

Law of the Sea and Exploitation of Marine Resources

Recently accelerated advances in marine technology enable countries to exploit ocean resources farther away from home and in waters deeper and previously inaccessible. Recognizing both the advantages of international cooperation in the scientific study of the sea and potential conflicts between competing uses of the sea, there is a growing consensus that a multi-national approach is needed to ensure the peaceful uses of the sea. The tradi-

tional "freedom of the seas" doctrine of international law, for example, can actually lead to intense rivalry over marine resources and the uses of the seas. Unlimited entry of new and potential exploiters will generally prevent any sort of voluntary restraints on exploitation and utilization of resources by present participants. Actually, the pressure will be to exploit as fast as possible!

Almost all participants in the current debate on the law of the sea, therefore, generally agree that some sort of international legal regime needs to be formulated which tends to limit entry and, at the same time, maximizes the net economic benefits from exploitation of marine resources so as to prevent misuse, overuse, destruction, waste, depletion and utter chaos. Two basic premises underlie this new thinking and approach to an effective international organization for controlling access to the ocean resources: (1) given the search for new food resources and the growing demand for energy and mineral resources, the marine technology which permits exploration and exploitation of marine resources is rapidly becoming available; (2) any major development of deep sea marine resources will require large capital investments and this in turn will require protection of access rights for continual, rational development and exploitation. With regard to these matters, two paramount issues are currently receiving considerable attention by the maritime nations of the world: What should be the seaward limit of the continental shelf? What resources are there beyond the continental shelf and who should control them?

Existing international law of the sea distinguishes between the internal waters and territorial sea of a nation; the high seas; the contiguous zone; the continental shelf; and the bed and subsoil of the deep seas, or the high seas beyond the continental shelf (see Figure 3-1).[11] While these global areas have no precise geographic references, it is important to distinguish between the rights of the United States *vis-à-vis* all other nations in these areas, and the division of authority between the federal government *vis-à-vis* the coastal states in areas acknowledged to come under the nation's jurisdiction.

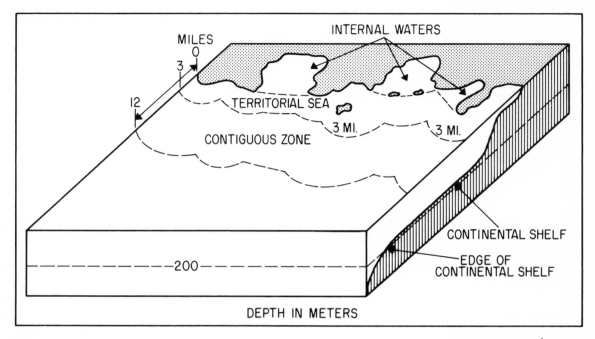

Figure 3-1. Relationship of internal waters, territorial sea, continguous zone and continental shelf.[1]

Internal Waters and Territorial Sea

Under the International Convention on the Territorial Sea and the Contiguous Zone, each nation's sovereignty extends beyond its land boundary "to a belt of sea adjacent to its coast, described as the territorial sea" and "to the air space over the territorial sea as well as to its bed and subsoil." The convention, however, does not specify the breadth of the territorial sea, and claims vary anywhere from 3 to 200 nautical miles or more. The United States recognizes no claim greater than 3 nautical miles as being sanctioned by international law. The coastal nation's "internal waters" constitute the rivers, lakes, and canals within the land area, the waters landward of the baseline of the territorial sea (generally the low waterline along the coast), and the waters landward of straight baseline across bays. Except for treaty limitations, each coastal nation maintains exclusive jurisdiction over its internal and territori-al waters and has exclusive, permanent access to the living and nonliving resources in these waters, on their beds, or in their subsoil.

High Seas

The International Convention on the High Seas defines "high seas" to include "all parts of the sea that are not included in the territorial sea or in the internal waters" of a nation. They are "open to all nations" and no nation "may validly purport to subject any part of them to its sovereignty." The freedoms of the seas may be exercised by both coastal and noncoastal nations; however, there are some restrictions or regulations imposed by international treaties and other international agreements. It is important to recognize that without a clear delimitation of the outer boundary of the territorial sea, jurisdictional conflicts are bound to occur between nations recognizing different limits, e.g., U.S. fishermen being seized for fishing in the "territorial waters" claimed by Peru which extend out to 200 miles from the coastline.

Contiguous Zone

The Convention on the Territorial Sea and the Contiguous Zone states that within a zone of the high seas, which may not extend beyond 12 miles from the baseline from which the breadth of the territorial sea is measured, each coastal nation may exercise the control necessary to prevent infringement of its customs, fiscal, immigration, or sanitary regulations and may penalize any such infringement committed within its territory or territorial sea. Although the convention seemingly does not extend national jurisdiction, *in toto,* in the contiguous zone, coastal nations have claimed permanent, exclusive access to living resources up to 12 miles from the coast, including the territorial sea. Thus, the United States prohibits foreign vessels from fishing within its 12-mile "exclusive fisheries zone" except under a bilateral agreement.

Continental Shelf

The International Convention on the Continental Shelf gives the coastal nation the sovereign and exclusive right to explore the shelf and exploit its natural resources. If the coastal nation does not itself engage in these activities, or make express claim to the continental shelf, no other party may do so without its permission.

The continental shelf as defined by the convention includes "the seabed and subsoil of the submarine areas adjacent to the coast but outside the area of the territorial sea, to a depth of 200 meters (656 feet) or, beyond that limit, to where the depth of the superjacent waters admits of the exploitation of the natural resources of the said areas" and "the seabed and subsoil of similar submarine areas adjacent to the coasts of islands." The natural resources to which the coastal nation is given permanent, exclusive access include "the mineral and other nonliving resources of the seabed and the subsoil together with living organisms. . .which, at the harvestable stage either are immobile on or under the seabed or are unable to move except in constant physical contact with the seabed or subsoil."

Division of Authority in the Coastal Zone

The coastal zone as viewed by the Commission on Marine Science, Engineering and Resources comprises ". . .seaward, the territorial sea of the United States and. . .landward, the tidal waters on the landward side of the low water mark along the coast,. . . ." Subject to the constitutional powers of Congress to regulate interstate and foreign commerce (including navigation), the states are primarily responsible for and authorized to manage the landward areas of the coastal zone. Until the passage of the Submerged Lands Act in 1953, the situation seaward was less clear. By virtue of this act, the coastal states, except for Texas and Florida, own the marine resources of the seabed and subsoil out to 3 nautical miles from their shores. Texas has authority over the resources out to 9 nautical miles from its coast, while Florida has authority out to 9 nautical miles from its Gulf coast.

Proposals for Rules Governing Exploitation of Marine Resources

The range of solutions in the current debate is tending to polarize between advocates of complete United Nations control of deep sea resources and those fearing the consequences of a U.N. takeover. Generally, it depends on which side of the technological "fence" a country is on. The "exploitability" criterion of the continental shelf convention provides, in effect, a "rubber boundary" that can be stretched by the increasing technological ability to exploit oil or mineral resources in waters deeper than 200 meters. Technologically advanced nations, therefore, will tend to resist narrowly restrictive limits.

Four alternative regimes have been suggested for coping with the problems raised by the potential development of marine resources beyond the continental shelf and slope.[12]

Division of the Sea Floor

This proposal would extend the coastal states' jurisdiction over the sea floor, seaward, until none remains free from some nation's control. Jurisdiction would involve only rights to resource exploitation. Aside from the unique problems posed by islands, this scheme is not likely to appeal to interests concerned with national security, e.g., the U.S. Navy. Although the division may be initially limited to exploitation rights, there is always the possibility that rights might be extended to cover, not only exploitation, but also use of the superjacent waters. If this occurs the mobility of naval vessels would certainly be impeded. However, the major reason for the unacceptability of this approach is that noncoastal nations will gain little or nothing from this approach. Furthermore, important nations in the political power struggle, e.g., the Soviet Union, may not find this allocation to their advantage.

Flag-Nation Approach

Under the flag-nation approach, the marine resources of the seabed beyond a coastal nation's jurisdiction would be open to exploitation and appropriation "under the laws of the flag of the discovering nation." The basic criticism, however, is that this approach does not ensure a "nonarbitrary" allocation of exclusive rights to the ocean seabed. There is no assurance that the resources will be used efficiently or even for productive purposes. Tenure and exclusive use would depend upon other nations' recognition of the claimant nation's right, and, unless they do, entrepreneurs may be unwilling to bear the risks of heavy capital investments given the ambiguous nature of their toehold.

International Registry Office

An improvement on the flag-nation approach is achieved by setting up an international registry office to record the claim of an entrepreneur through his nation. The international authority would have the power to designate "zones of special jurisdiction" which are subject to the laws of the claimant nation, but would be limited with respect to area, purpose and duration of occupancy, and time to actually engage in development. The advantages of this regime are that: (1) it provides for more security than just a flag-nation regime; and (2) it tends to avoid overexploitation of the marine resources as would occur under a "common property" regime. It is argued, however, that this regime would inevitably operate on a "first-come, first-served" basis and, given the "use it or lose it" performance requirement, would tend to induce overexploitation rather than lessen it.

International Authority

Through an extension of the above scheme, an international authority would establish a market for the licenses, or permits, guaranteeing exclusive rights to exploit the resources of the deep sea. This, in effect, gives the international authority jurisdiction, or ownership, over exploitation rights to deep sea floor resources. The argument is that by requiring nations to bid for rights, this would maximize the "economic rent," or yield, of the marine resources of the seabed and would also grant the rights to efficient exploiters. The argument is further made that this regime is analogous to the U.S. Department of the Interior policy of granting rights in offshore oil drilling. However, there would still remain problems of reconciling "competing" uses of the seabed and valuing both these, and the "external" benefits and costs to the international community—which obviously is easier for a national regime, compared to an international regime, to cope with.

As to the extent of jurisdiction over the continental shelf, almost all engaged in the debate agree that some precise limits should be drawn, but no definite agreement has yet been reached as to the extent of those limits. Authoritative public and private groups have advanced three recent proposals.

Narrow Limits

Participants of the Thirty-third American Assembly met in May 1968 at Arden House, New York, to discuss a compendium of papers on the "Uses of the Seas." They endorsed the proposal of a territorial sea "as narrow as possible" and "the shelf as redefined should be narrow, not wide, preferably not beyond the 200 meter depth;"[13] and, further, that an international regime be established and have jurisdiction over exploitation rights beyond this limit.

Wide Limits

The Committee on Petroleum Resources Under the Ocean Floor of the National Petroleum Council (NPC), created in response to a request by the Interior Department for assistance in policy-formation, recommended that U.S. title should be asserted to the "full extent" of the continental land mass down to the ocean floor.[14] This may range down to depths anywhere from 5,000 to 15,000 feet depending upon the topography of the continental slope and rise and would thus give some nations jurisdiction over larger areas than others. They also recommended that the resources of the deep ocean floor not be turned over to an international licensing authority but, rather, that exploration and recovery be subject to international agreement on codes of conduct and negotiation among nations.

Redefined Limits

The Report of the Commission on Marine Science, Engineering and Resources rejected the NPC proposal as being "contrary to the best interests of the United States," while stating that the nation's "security and world peace are best served by the narrowest possible definition of the continental shelf for purposes of mineral resources development."[15] The Commission recommended two limits:

1. Each coastal nation's juridical continental shelf would be fixed at the 200-meter isobath, or 50 nautical miles from shore, whichever gives the greater area. The rationale for these limits is that 200 meters is the average depth of the outer edges of the world's continental shelves and 50 nautical miles is their average width;

2. An "intermediate zone" should be created, encompassing the bed and subsoil of the deep seas out to the 2,500-meter isobath, or 100 nautical miles from shore whichever gives the greater area. Again, 2,500 meters is considered the average depth of the world's geological continental slopes and 100 miles the average width of the shelves and slopes.

Within the zone straddling these two limits, exclusive rights to explore and exploit can be granted only by the coastal nation subject to criteria consistent with an international regime for governing the deep ocean floor. For example, the United States might continue to grant leases through bidding in the intermediate zone, but a part of the revenues would be paid into an International Fund "for purposes that the international community agrees will promote the common welfare." By using a dual criteria in defining limits of national jurisdiction, this scheme provides for a greater degree of equity than occurs by natural topography by giving something extra to those nations having extremely narrow continental shelves. It also tends to minimize the possibility of a headlong race for claims to seabed resources and provides more scope for reconciling competing uses of the seas.

For the region beyond national limits, the Commission recommends an international registry authority to register claims to explore or exploit mineral resources on a "first-come, first-registered" basis. Payments for leases to exploit particular mineral resources in a particular area of the deep seas collected by the authority would be turned over to an International Fund "to be expended for such purposes as financing marine scientific activity and resources exploration and development, particularly food-from-the-sea programs, and aiding the developing countries through the World Bank U.N. Development Program and other international development agencies."

United Nations Considerations

The United Nations has also been considering the question of utilization of the resources of the seabed and subsoil beyond the limits of national jurisdiction.[16] In December 1967 the U.N. General Assembly adopted the resolution:

> Examination of the Question of the Reservation Exclusively for Peaceful Purposes of the Sea-Bed and the Ocean Floor, and the Subsoil Thereof, Underlying the High Seas Beyond the Limits of Present National Jurisdiction, and the Uses of Their Resources in the Interests of Mankind,

whereupon an *ad hoc* Committee of 35 countries was established, including the United States, to study the item and present its report to the twenty-third session of the General Assembly. Following submission of the report, the General Assembly in December 1968 adopted a resolution to establish a Standing Committee on the Peaceful Uses of the Sea-Bed and the Ocean Floor Beyond the Limits of National Jurisdiction, composed of 42 countries, to study further ways and means of promoting international cooperation in the exploration and use of the seabed and subsoil of the ocean floor, taking into account the development of technology and the economic considerations of such utilization and to study the elaboration of a legal regime which would ensure that control and use of deep sea resources would benefit all mankind.

References

1. Report of the Commission on Marine Science, Engineering and Resources. *Our Nation and the Sea—A Plan for National Action*. Washington: U.S. Government Printing Office, January 1969, 1.

2. Marine Resources and Engineering Development Act of 1966 (Public Law 89-454, June 17, 1966). Cited in Appendix B-2. The third report of the President to the Congress on Marine Resources and Engineering Development. *Marine Science Affairs—A Year of Broadened Participation*. Washington: U.S. Government Printing Office, January 1969, 223.

3. National Sea Grant College and Program Act of 1966 (Public Law 89-688, October 15, 1966). Cited in Appendix B-3, 228.

4. The first report of the President to the Congress on Marine Resources and Engineering Development. *Marine Science Affairs—A Year of Transition*. Washington: U.S. Government Printing Office, February 1967, III.

5. The second report of the President to the Congress on Marine Resources and Engineering Development. *Marine Science Affairs—A Year of Plans and Progress*. Washington: U.S. Government Printing Office, February 1968, 15-18.

6. The third report, 8-10.

7. Christy, Francis T. Jr. and Anthony Scott. *The Common Wealth in Ocean Fisheries*. Baltimore: The Johns Hopkins Press, 1965, 67-70.

8. Commission on Marine Science, Engineering and Resources, Panel Reports, Vol. 2. *Keys to Oceanic Development*. Washington: U.S. Government Printing Office, 1969, V-34-42.

9. The second report, 39-40.

10. D. L. Alverson and E. A. Schaefers. "Methods of Search and Capture in Ocean Fisheries," *Exploiting the Ocean*. Washington: Transactions of the Second Annual Conference and Exhibit of Marine Technology Society, 1966, 319-335.

11. Commission on Marine Science, Engineering and Resources, Panel Reports, Vol. 3. *Marine Resources and Legal-Political Arrangements for Their Development*. Washington: U.S. Government Printing Office, 1969, VIII-10-15.

12. Cf. Christy, Jr., Francis T. "Economic Criteria for Rules Governing Exploitation of Deep Sea Minerals," *The International Lawyer.* (January 1968), 224-42. (Reprint No. 72, March 1968, Resources for the Future, Inc.)

13. Report of the Thirty-third American Assembly. "Uses of the Seas." (May 2-5, 1968, Harriman, New York), 6. (See also: Edmund A. Gullion, editor. The American Assembly, "Introduction: New Horizons at Sea," *Uses of the Seas.* Englewood Cliffs, N.J.: Prentice-Hall, Inc., 1968, 1-16.)

14. Cf. *Petroleum Resources Under the Ocean Floor.* Washington: National Petroleum Council, 1969.

15. Report of the Commission, 144-45.

16. Cf. "Appendix C-1—Resolutions Adopted by United Nations General Assembly." Cited in the third report, 233-241.

part 4: giulio venezian

Non-linear Spin-up

Dr. Giulio Venezian received his B. Eng. in engineering physics from McGill University in 1960 and his Ph.D. in engineering science from the California Institute of Technology in 1965.

After completing his doctoral work, Dr. Venezian continued at the California Institute of Technology as Research Fellow until 1968.

Dr. Venezian is now assistant professor of ocean engineering at the University of Hawaii.

4.

non-linear spin-up

Introduction

An introduction to spin-up was given in Chapter 18 of *Topics in Ocean Engineering, Volume 1*. Wherever it is needed, material will be drawn from that chapter; but, to keep this article self-contained, some material will be repeated, although not in detail.

Three distinct topics will be discussed here. One is the application of Wedemeyer's theory to the decay of transients in cases where the final state is not rigid body rotation. The second topic is an analysis of the structure of the moving shear discontinuities that occur in the nonlinear transients under consideration. The third topic is a formulation for more general spin-up geometries. These topics are disconnected, and can be regarded as three separate appendices to the above reference.

Cylinder With Top and Bottom Rotating at Different Rates

This section deals with the problem of a cylinder of radius a and height h filled with a fluid of low viscosity which initially rotates at an angular speed ω about the axis of the cylinder. At time t = 0 the angular velocity of the bottom boundary is increased to a speed Ω_B, while that of the top boundary is increased to Ω_T. Of interest are the final flow as well as the details of the transient.

In Reference 1, it was shown that the flow in the interior is governed by the equations

$$\frac{\partial V}{\partial t} + \frac{U}{r} \frac{\partial}{\partial r} (rV) = 0 \tag{4-1}$$

$$\frac{\partial U}{\partial z} = \frac{\partial V}{\partial z} = 0 \tag{4-2}$$

$$\frac{1}{r} \frac{\partial}{\partial r} (rU) + \frac{\partial W}{\partial z} = 0 \tag{4-3}$$

(See Equations 18-40 to 18-42 in Reference 1.)

In addition, it was shown that at the boundaries, the flow into the boundary layer is connected to the vorticity of the interior flow. If the relationship can be assumed to be linear, then the values of W at the top (z = h/2) and at the bottom (z = -h/2) are

$$W_T = -\left(\frac{\nu}{\Omega_T}\right)^{\frac{1}{2}} \frac{1}{2r} \frac{\partial}{\partial r} (rV - r^2\Omega_T) \tag{4-4}$$

$$W_B = \left(\frac{\nu}{\Omega_B}\right)^{\frac{1}{2}} \frac{1}{2r} \frac{\partial}{\partial r} (rV - r^2 \Omega_B) \tag{4-5}$$

(See Equations 18-31, 18-32 and 18-44 in Reference 1.)

From Equations 4-2 and 4-3, it follows that the dependence of W on z is at most linear. It follows that

$$W = \left(½ + \frac{z}{h}\right) W_T + \left(½ - \frac{z}{h}\right) W_B$$

(4-6)

and, in view of Equation 4-3,

$$\frac{1}{r} \frac{\partial}{\partial r} (rU) = \frac{W_B - W_T}{h}$$

(4-7)

This equation can be integrated directly to give

$$U = \frac{\nu^{½}}{2h} (\Omega_B^{½} + \Omega_T^{½}) \left(\frac{V}{\Omega_B^{½} \Omega_T^{½}} - r\right)$$

(4-8)

When the steady state is reached, U should be zero, and hence

$$V = \Omega_B^{½} \Omega_T^{½} r$$

(4-9)

It should be noted that in arriving at this result, the boundary layer equations at the two horizontal surfaces were linearized about the velocities of the corresponding surfaces. If they had been linearized about a single angular velocity the result would have been

$$V = ½ (\Omega_B + \Omega_T) r$$

(4-10)

The problem of the steady state flow between two infinite plates rotating at different rates has been discussed by Batchelor,[2] Stewartson[3] and Lance and Rogers.[4]

Batchelor discussed the problem qualitatively, arguing that the two-plate solution can be obtained from the one-plate solution by matching the axial velocity in the interior. This is, in fact, the procedure used in arriving at Equation 4-6. Stewartson expressed some reservations—in particular, for the case where the plates are rotating in opposite directions. Lance and Rogers performed numerical calculations which showed that in fact the single plate solutions can be combined when

the plates rotate in the same direction but not necessarily when they rotate in opposition.

In an earlier paper, Rogers and Lance[5] had performed numerical calculations for the single disc problem, and from their results it is possible to obtain a curve of Ω/Ω_B against Ω_T/Ω_B. Such a curve was obtained by Greenspan[6] and appears in Figure 4-1. This figure also shows the curves corresponding to Equations 4-9 and 4-10. It appears that the curve obtained from the exact numerical calculations lies approximately halfway between the other two, and it is therefore tempting to compare it with the average of the latter two curves

$$\Omega = \left(\frac{\Omega_B^{½} + \Omega_T^{½}}{2}\right)^2$$

(4-11)

This curve coincides almost exactly with the exact curve except for a small portion near $\Omega_T/\Omega_B = 0$, which is shown in Figure 4-1 as a dotted line.

Equation 4-11 is therefore taken as the correct expression for the steady state angular velocity of the interior, and Equation 4-8 is modified to read

$$U = \frac{\nu^{½}}{\Omega^{½} h} (V - \Omega r)$$

(4-12)

where Ω is given by Equation 4-11.

Transient State

The details of the transient state can now be worked out. Combining Equations 4-1 and 4-12,

$$\frac{\partial V}{\partial t} + E^{½} (V - \Omega r) \frac{1}{r} \frac{\partial}{\partial r} (rV) = 0$$

(4-13)

which was termed Wedemeyer's equation in Reference 1.

Here $E = \nu/\Omega h^2$ is the Ekman number, which—throughout this discussion—is assumed to be very small.

Equation 4-13 has to be solved with the initial condition $V = \omega r$ at $t = 0$ and subject to the condition that $U = 0$ at the outer radius of the cylinder, $r = a$. The solution of this problem is given in

Reference 1 (Equations 18-49, 18-51) and consists of two pieces

$$rV = \Omega r^2 \bigg/ \left[1 + \frac{\Omega - \omega}{\omega} e^{-2\Omega E^{\frac{1}{2}} t} \right]$$

(4-14)

for $r < r_o$, and

$$rV =$$
$$(\Omega r^2 - \Omega a^2 e^{-2\Omega E^{\frac{1}{2}} t}) \bigg/ (1 - e^{-2\Omega E^{\frac{1}{2}} t})$$

(4-15)

for $a > r > r_o$.

In these expressions

$$r_o = a \left[\frac{\omega}{\Omega} + \frac{\Omega - \omega}{\Omega} e^{-2\Omega E^{\frac{1}{2}} t} \right]^{\frac{1}{2}}$$

(4-16)

Thus, the flow is very similar to spin-up in a rigid cylinder. The velocity components V and U correspond to those in a cylinder spinning up to an angular speed Ω.

The principal difference is in W, which is no longer antisymmetric in z. Because of this, it is no longer true that the boundary layers in $r < r_o$ suck fluid and those in $r > r_o$ expel fluid.

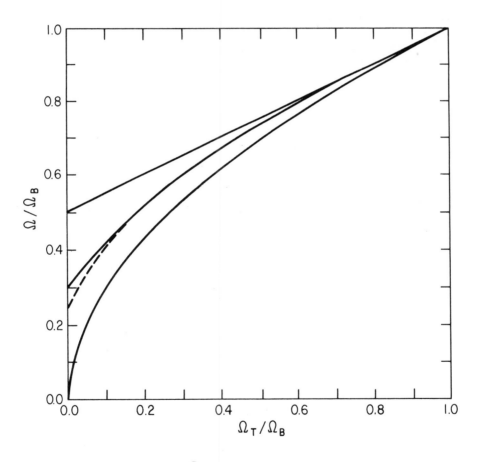

Figure 4-1. Graph of Ω/Ω_B vs. Ω_T/Ω_B. Top curve obtained from Equation 4-10, bottom curve from Equation 4-9 and center curve from References 5 and 6. Dotted curve represents Equation 4-11.

As an example, consider the case $\omega = 0$. Equation 4-14 states that for $r < r_o$, $V = 0$, and from Equation 4-16

$$r_o = ae^{-\Omega E^{1/2}t} \qquad (4\text{-}17)$$

In the region outside r_o,

$$V = \Omega r (1 - r_o^2/r^2)\Big/(1 - r_o^2/a^2) \qquad (4\text{-}18)$$

so that the flow appears to be identical to the one that obtains when a cylinder is spun up from $\omega = 0$ to a final angular velocity Ω.

The corresponding expressions for U are

$$U = -E^{1/2}\Omega r \qquad (4\text{-}19)$$

for $r < r_o$ and

$$U = E^{1/2}\Omega e^{-2\Omega E^{1/2}t} \frac{(r - \frac{a^2}{r})}{1 - e^{-2\Omega E^{1/2}t}} \qquad (4\text{-}20)$$

Figure 4-2. Typical flows for spin-up from rest for $\Omega_T/\Omega_B = 0.5$. (a) is for $t < 1.3$, (b) for $t > 1.3$. Dotted lines indicate the position of the shear discontinuity.

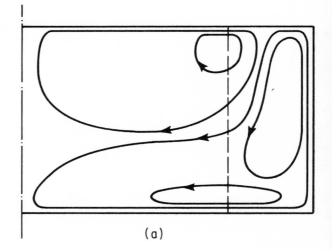

(a)

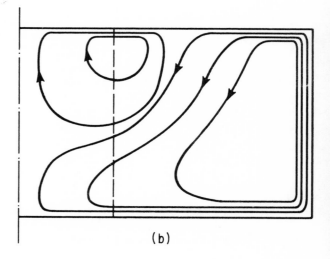

(b)

for $a > r > r_o$.

The expressions for W_T and W_B are

$$W_T = (\nu \Omega_T)^{1/2} \ , \ r < r_o$$

$$= (\nu \Omega_T)^{1/2} \left(1 - \frac{\Omega/\Omega_T}{1 - e^{-2\Omega E^{1/2} t}}\right) , \ r > r_o$$

$$(4\text{-}21)$$

$$W_B = - (\nu \Omega_B)^{1/2} \ , \ r < r_o$$

$$= - (\nu \Omega_B)^{1/2} \left(1 - \frac{\Omega/\Omega_B}{1 - e^{-2\Omega E^{1/2} t}}\right) , \ r > r_o$$

$$(4\text{-}22)$$

It can be seen from these expressions that in the region $r > r_o$ W_B changes sign as time progresses. It is initially positive but becomes negative at large times. Thus, for short times, in the region $r < r_o$ fluid is pumped into the walls and drawn in radially and it is expelled by the walls in the region $r > r_o$ completing the flow. For larger times, however, the faster plate sucks fluid throughout its length. The flow patterns are shown schematically in Figure 4-2.

Taking $\Omega_T / \Omega_B = 0.5$, it follows from Equation 4-11 that $\Omega / \Omega_B = 0.728$. Thus, W_B changes sign when

$$\Omega/\Omega_B = 1 - e^{-2\Omega E^{1/2} t}$$

that is, when $e^{-2\Omega E^{1/2} t} = 0.272$, or $2\Omega E^{1/2} t = 1.3$.

Other examples can be worked out in a similar fashion. All the flows exhibit a moving shear discontinuity at $r = r_o$ which eventually stops at a radius such that its initial angular momentum density is the correct one for the final flow. Spin-up is achieved by pumping at the horizontal walls, which suck fluid if the local vorticity of the interior flow is less than that of the wall and expel fluid if the opposite is the case.

The role of the side walls is only secondary and was discussed in Reference 1. To the degree of approximation used here, the angular velocity of

the side wall is immaterial. Its only function is to prevent fluid from crossing it, which—because of Equation 4-12—automatically implies that $V = \Omega a$ at the wall. If the velocity of the wall is other than this value, a boundary layer must accomodate the discontinuity.

Structure of the Shear Discontinuity

In the flows discussed thus far, different expressions were obtained for the velocity components in the regions $r < r_o$ and $r > r_o$. While the velocity V is continuous across this surface, its derivative $\partial V / \partial r$ is not. Thus, there is a discontinuity in shear. The surface could also be termed a velocity discontinuity because W is discontinuous across $r = r_o$; but, since V constitutes the primary part of the flow, the term "shear discontinuity" will be used.

A discontinuity of this kind implies that the effects of viscosity, which are neglected in arriving at the interior equations, are not negligible in the neighborhood of the discontinuity. In order that the flow field obtained be a valid one, it is necessary to show that the discontinous portions can be joined smoothly by a boundary layer—in this case, a moving layer.

To examine the layer, an equation must be obtained to replace Equation 4-13, in which viscous stresses were neglected. The effect of the lateral stresses must be taken into account. Jacobs[8] has shown that it is possible to do this by simply retaining the lateral stress term from the momentum equation, without altering the expression for U given in Equation 4-12, provided the scale of lateral shear is greater than the scale of the top and bottom layers.

The resulting equation, which was given by Wedemeyer,[7] is

$$\frac{\partial V}{\partial t} + E^{1/2} (V - \Omega r) \frac{1}{r} \frac{\partial}{\partial r} (rV) = \nu \left(\frac{\partial^2 V}{\partial r^2} + \frac{1}{r} \frac{\partial V}{\partial r} - \frac{V}{r^2}\right)$$

$$(4\text{-}23)$$

The expression is simplified if the angular momentum density, $G = rV$ is used as the dependent variable. The equation then becomes

$$\frac{\partial G}{\partial t} + 2E^{1/2} (G - \Omega r^2) \frac{\partial G}{\partial (r^2)} = 4\nu r^2 \frac{\partial^2 G}{\partial (r^2)^2}$$

$$(4\text{-}24)$$

Now the analysis will be restricted to the case of spin-up from rest, so that G should satisfy the following requirements:

$$G \to 0 \quad \text{for} \quad r << r_o \qquad (4\text{-}25)$$

$$G \to \frac{\Omega (r^2 - r_o^2)}{1 - r_o^2 / a^2} \quad \text{for} \quad r >> r_o \qquad (4\text{-}26)$$

This would make V agree with the values previously obtained. In these expressions r_o is the location of the shear discontinuity; and, as given by Equation 4-17,

$$r_o = a \, e^{-\Omega E^{1/2} t}$$

Since the principal interest here is in the structure of the flow in the vicinity of r_o it is desirable to define a new set of coordinates in such a way as to make the position of the shear discontinuity a fixed value of one of these coordinates. Accordingly,

$$\xi = r^2 e^{2\Omega E^{1/2} t} / a^2 \qquad (4\text{-}27)$$

$$\eta = e^{2\Omega E^{1/2} t} - 1 \qquad (4\text{-}28)$$

Then

$$\frac{\partial}{\partial (r^2)} = \frac{\eta + 1}{a^2} \frac{\partial}{\partial \xi} \qquad (4\text{-}29)$$

and

$$\frac{\partial}{\partial t} = 2\Omega E^{1/2} \left(\xi \frac{\partial}{\partial \xi} + (\eta + 1) \frac{\partial}{\partial \eta} \right) \qquad (4\text{-}30)$$

With these substitutions, Equation 4-24 becomes

$$\frac{\partial G}{\partial \eta} + \frac{1}{\Omega a^2} G \frac{\partial G}{\partial \xi} = \frac{2\nu}{\Omega E^{1/2}} \frac{\xi}{a^2} \frac{\partial^2 G}{\partial \xi^2} \qquad (4\text{-}31)$$

and the position of the shear discontinuity is now at $\xi = 1$.

It is clear that the viscous term will only be significant if

$$\frac{\partial}{\partial \xi} \sim \frac{a}{h} E^{-1/4}$$

Accordingly, it is convenient to rescale ξ setting

$$\xi = \left(1 + \frac{2hE^{1/4}}{a} \zeta \right) \qquad (4\text{-}32)$$

and at the same time, let

$$g = \frac{G}{2hE^{1/4} \Omega a} \qquad (4\text{-}33)$$

Then

$$\frac{\partial g}{\partial \eta} + g \frac{\partial g}{\partial \zeta} = \frac{1}{2} \left(1 + \frac{2hE^{1/4}}{a} \zeta \right) \frac{\partial^2 g}{\partial \zeta^2} \qquad (4\text{-}34)$$

This equation is almost identical to Burgers' equation[9] for which a solution is known.[10,11] To make the equations identical, it is necessary to drop the term

$$\frac{hE^{1/4}}{a} \zeta \frac{\partial^2 g}{\partial \zeta^2}$$

Now, because ζ is a "stretched" coordinate, its range is $-\infty$ to ∞. Thus, even though the coefficient in front is small, the term ultimately becomes large. Nevertheless, the term will be in error only far away from the shear discontinuity, where $\partial^2 g / \partial \zeta^2$ is expected to be negligible anyway. This justifies replacing Equation 4-34 by

$$\frac{\partial g}{\partial \eta} + g \frac{\partial g}{\partial \zeta} = \frac{1}{2} \frac{\partial^2 g}{\partial \zeta^2} \qquad (4\text{-}35)$$

This equation must be solved subject to

$$g \to 0 \quad \text{for} \quad \zeta \to -\infty \qquad (4\text{-}36)$$

and

$$g \rightarrow \zeta/\eta \quad \text{for} \quad \zeta \rightarrow \infty \qquad (4\text{-}37)$$

which are simply Equations 4-25 and 4-26, restated in terms of the new variables.

References 10 and 11 show that the substitution

$$g = -\frac{1}{\psi} \frac{\partial \psi}{\partial \zeta} \qquad (4\text{-}38)$$

will transform Equation 4-35 into

$$\frac{\partial \psi}{\partial \eta} = \frac{1}{2} \frac{\partial^2 \psi}{\partial \zeta^2} \qquad (4\text{-}39)$$

The "initial" conditions at $\eta = 0$ are $g = 0$ for $\zeta < 0$ and $g = \infty$ for $\zeta > 0$. In terms of ψ these are $\psi = 1$ for $\zeta < 0$ and $\psi = 0$ for $\zeta > 0$.

Now, the solution of Equation 4-39 is

$$\psi = \frac{1}{\sqrt{2\pi\eta}} \int_{-\infty}^{\infty} e^{-(\zeta - \zeta')^2/2\eta} \psi_{(\eta=0)} \, d\zeta'$$

$$= \frac{1}{\sqrt{2\pi\eta}} \int_{-\infty}^{0} e^{-(\zeta - \zeta')^2/2\eta} \, d\zeta'$$

$$= \frac{1}{2} \, \text{erfc}\left(\frac{\zeta}{\sqrt{2\eta}}\right) \qquad (4\text{-}40)$$

It follows that

$$g = \frac{2 \, e^{-\zeta^2/2\eta}}{\sqrt{2\pi\eta} \, \text{erfc}\left(\frac{\zeta}{\sqrt{2\eta}}\right)} \qquad (4\text{-}41)$$

Now for large positive values of its argument,

$$\text{erfc} \quad u \rightarrow e^{-u^2}/ u \sqrt{\pi}$$

and, therefore, for $\zeta \rightarrow \infty$

$$g \rightarrow \zeta/\eta$$

while for large negative values of u

$$\text{erfc} \quad u \rightarrow 2 + e^{-u^2}/u\sqrt{\pi}$$

so that for $\zeta \rightarrow -\infty$

$$g \rightarrow \frac{1}{\sqrt{2\pi\eta}} \, e^{-\zeta^2/2\eta}$$

These expressions are in agreement with the requirements on g given in Equations 4-36 and 4-37, so that the expression for g given in Equation 4-41 matches smoothly the two solutions across $r = r_0$.

The moving shear discontinuity is thus a layer of thickness proportional to $E^{1/4}$. In similar linear problems, Stewartson[12] showed that there are layers with a double structure of thicknesses proportional to $E^{1/4}$ and $E^{1/3}$. What apparently happens in the nonlinear case is that the $E^{1/4}$ layer is a moving layer, while the $E^{1/3}$ layer (which is required to satisfy W=0 at the wall) remains attached to the wall. Thus, the nonlinearity separates the double layer.

Formulation of Problem for Asymmetric Geometries

In this section a formulation of the spin-up problem will be developed for cylindrical containers having a cross-section other than circular.

The flow can be expected to consist of two components—a primary two-dimensional flow having no dependence on z, the coordinate parallel to the axis of rotation; and a secondary flow, also essentially two-dimensional driven by the difference between the vorticity in the primary interior flow and that of the top and bottom boundaries.

Since the container will be rotating with an angular velocity Ω and its shape will appear constant in a frame of reference that rotates at that speed, it is convenient to formulate the problem with reference to such a frame. The z-component of the vorticity equation is then

$$\frac{\partial \zeta}{\partial t} + \vec{q} \cdot \nabla \zeta = \vec{\omega} \cdot \nabla w + 2\Omega \frac{\partial w}{\partial z} \qquad (4\text{-}42)$$

In this expression $\vec{q}$ is the fluid velocity (referred to the rotating frame) and its components, in Cartesian coordinates, are u, v, w. The curl of $\vec{q}$

is the vorticity, denoted by $\vec{\omega}$ and having components ξ, η, ζ.

Because of the assumptions of two-dimensionality, the primary flow is described in terms of a stream function ψ as follows:

$$u = -\frac{\partial \psi}{\partial y} \, , \; v = \frac{\partial \psi}{\partial x} \, , \; w = 0$$

The secondary flow, however, has to provide the inflow into the boundary layers. Thus, while u and v are not functions of z, w is non-zero; and, to satisfy the equation of continuity, w must be a linear function of z. Accordingly, a general representation of the secondary flow is

$$u = \frac{\partial \phi}{\partial x} \, , \; v = \frac{\partial \phi}{\partial y} \, , \; w = -z\nabla^2 \phi$$

This representation satisfies the condition $\nabla \cdot \vec{q} = 0$.

Combining the two representations, the total flow is given by

$$u = \frac{\partial \phi}{\partial x} - \frac{\partial \psi}{\partial y} \tag{4-43}$$

$$v = \frac{\partial \phi}{\partial y} + \frac{\partial \psi}{\partial x} \tag{4-44}$$

$$w = -z\nabla^2 \phi \tag{4-45}$$

and the corresponding vorticity components are

$$\xi = -z\nabla^2 \frac{\partial \phi}{\partial y} \tag{4-46}$$

$$\eta = +z\nabla^2 \frac{\partial \phi}{\partial x} \tag{4-47}$$

$$\zeta = \nabla^2 \psi \tag{4-48}$$

Equation 4-42 thus becomes

$$\frac{\partial \zeta}{\partial t} + \nabla \phi \cdot \nabla \zeta + \vec{k} \cdot (\nabla \psi \cdot \nabla \zeta) =$$

$$\vec{\omega} \cdot \nabla w + 2\Omega \frac{\partial w}{\partial z} \tag{4-49}$$

Now, since the secondary flow is driven by the boundary layers, it must be of order $E^{1/2}$ compared to the primary flow. The time variations, which occur because of the secondary flow will also be slow, so that time derivatives are also of order $E^{1/2}$. It follows that all the terms in the above equation are of order $E^{1/2}$ or smaller compared to $\vec{k} \cdot (\nabla \psi \cdot \nabla \zeta)$, so that this term in itself must be zero. This is merely a statement that the primary flow should be an acceptable steady flow by itself. In terms of the derivatives of ψ and ζ,

$$\frac{\partial \psi}{\partial x} \frac{\partial \zeta}{\partial y} - \frac{\partial \psi}{\partial y} \frac{\partial \zeta}{\partial x} = 0 \tag{4-50}$$

so that ζ must be a function of ψ (and not of x and y individually). Thus,

$$\zeta = F(\psi, t) \tag{4-51}$$

The term $\vec{\omega} \cdot \nabla w$ contains only one product of order $E^{1/2}$, namely $\zeta \, \partial w/\partial z$, while the other two terms are smaller. Thus, only this larger term is retained with the result

$$\frac{\partial \zeta}{\partial t} + \nabla \phi \cdot \nabla \zeta + \zeta \nabla^2 \phi = -2\Omega \nabla^2 \phi \tag{4-52}$$

What remains to be done is to provide a relation between ϕ and ζ i.e. between the primary and secondary flows. The relationship is provided by the boundary layer equations

$$w(-h/2) = -w(h/2) = \frac{1}{2} (\nu/\Omega)^{\frac{1}{2}} \zeta \tag{4-53}$$

(See Equations 18-31, 18-32 in Reference 1 and also Equations 4-4 and 4-5 above.)

It follows that

$$w = -zE^{\frac{1}{2}} \zeta$$

or

$$\zeta = E^{-\frac{1}{2}}\nabla^2 \phi \qquad (4-54)$$

Equations 4-52 and 4-54 describe the flow and are equivalent to Wedemeyer's equation for axially symmetric flow.

Substituting Equation 4-51 into Equation 4-52, the result is

$$\frac{\partial F}{\partial t} + \frac{\partial F}{\partial \psi}\left(\frac{\partial \psi}{\partial t} + \nabla\phi \cdot \nabla\psi\right) + (F + 2\Omega)\nabla^2\phi = 0 \qquad (4-55)$$

Now, F is a function of ψ and t only, and by Equation 4-54 the same applies to $\nabla^2\phi$. It follows that the coefficient of the second term must also be a function of ψ and t only.

$$\frac{\partial \psi}{\partial t} + \nabla\phi \cdot \nabla\psi = G(\psi, t) \qquad (4-56)$$

It is now possible to express Equations 4-55 and 4-56 as directional derivatives.

$$\frac{dF}{ds} = -E^{\frac{1}{2}} F(F + 2\Omega) \qquad (4-57)$$

$$\frac{dt}{ds} = 1 \qquad (4-58)$$

$$\frac{d\psi}{ds} = G(\psi, s) \qquad (4-59)$$

From Equations 4-57, 4-58 and 4-59, it follows that

$$\zeta = \frac{2\Omega h(\psi, t)}{e^{2\Omega E^{\frac{1}{2}}t} - h(\psi, t)} \qquad (4-60)$$

will satisfy the equations of motion for any function $h(\psi,t)$ that satisfies

$$\frac{\partial h}{\partial t} + \nabla\phi \cdot \nabla h = 0 \qquad (4-61)$$

Thus, the equations of motion have been integrated in part. In the case of a circular cylinder, it

turns out that h is in fact a constant, so that the remaining integrations of Equations 4-48 and 4-54 can be carried out readily. Unfortunately, this does not appear to be the case for other shapes of cylinders. The formulation thus has to stop at this point.

The following points should be noted—the function h is convected by the secondary flow and therefore remains constant relative to a point moving with a velocity $\nabla\phi$. Thus, even though h will get "moved around," the container, it will not grow in time. Equation 4-60 then shows that the vorticity excess ζ will decay in a time of order $1/2\Omega E^{1/2}$.

It is interesting also that the linear case is easily recovered from Equation 4-52, which—upon dropping the nonlinear terms—becomes

$$\frac{\partial \zeta}{\partial t} + 2\Omega E^{\frac{1}{2}}\zeta = 0 \qquad (4-62)$$

Thus,

$$\zeta = \zeta_o(x, y) e^{-2\Omega E^{\frac{1}{2}}t} \qquad (4-63)$$

It is possible that in the nonlinear case, a non-steady primary flow is needed (i.e. one which changes in a time scale of order $1/\Omega$ in addition to the time scale $1/\Omega E^{1/2}$ considered here.) Further work is required to resolve this point.

References

1. Venezian, G. "Spin-up of a Contained Fluid," *Topics in Ocean Engineering, Volume 1.* Houston: Gulf Publishing Co., 1969, 212-223.

2. Batchelor, G. K. "Note on a Class of Solutions of the Navier-Stokes Equations Representing Steady Rotationally-symmetric Flow," *Quart. J. Math. App. Mech.*, 4 (1951), 29.

3. Stewartson, K. "On the Flow between Two Rotating Coaxial Disks," *Proc. Camb. Phil. Soc.*, 49 (1953), 323.

4. Lance, G. N. and M. H. Rogers. "The Axially Symmetric Flow of a Viscous Fluid between Two Infinite Rotating Disks," *Proc. Roy. Soc.*, A266 (1962), 109.

5. Rogers, M. H. and G. N. Lance. "The Rotationally Symmetric Flow of a Viscous Fluid in the Presence of an Infinite Rotating Disk," *J. Fluid Mech.*, 7 (1960), 617.

6. Greenspan, H. P. *The Theory of Rotating Fluids.* Cambridge University Press, 1968, 146.

7. Wedemeyer, E. H. "The Unsteady Flow within a Spinning Cylinder," *J. Fluid Mech.*, 20 (1964), 383.

8. Jacobs, S. J. "The Taylor Column Problem," *J. Fluid Mech.*, 20 (1964), 581.

9. Burgers, J. "A Mathematical Model Illustrating the Theory of Turbulence," *Advances in Applied Mechanics.* New York: Academic Press, 1948.

10. Cole, J. D. "On a Quasi-linear Parabolic Equation Occurring in Aerodynamics," *Q. App. Math.,* 9 (1951), 225.

11. Hopf, E. "The Partial Differential Equation $u_t + uu_x = u_{xx}$," *Comm. Pure Appl. Math.,* 3 (1950), 201.

12. Stewartson, K. "On Almost Rigid Rotations," *J. Fluid. Mech.,* 3 (1957), 17.

part 5: george carrier

Tsunamis and Wave Run-up

Hurricanes

The Midlatitude Ocean Circulation

A Gordon McKay professor of mechanical engineering, Dr. George F. Carrier has been at Harvard University since 1952. Before joining the faculty there, he was at Brown University as assistant professor 1946-47, associate professor 1947-48 and professor 1948-52.

As well as being an educator and author, Dr. Carrier has served on assorted committees, panels and boards. He has been with the Committee on Atmospheric Sciences, Nat'l Academy of Sciences; Corporation, Woods Hole Oceanographic Institution; U.S. Nat'l Committee on Theoretical and Applied Mechanics, of which he was chairman 1968-69; Subcommittee on Theory Project Committee, Nat'l Academy of Sciences; Education and Manpower Panel, Comm. on Atmospheric Science, Nat'l Academy of Sciences; Committee on Graduate and Postdoctoral Education, Nat'l Research Council; Visiting Committee, Dept. of Mechanics, Lehigh University; and Council for the Engineering College, Cornell University, 1968-69.

His affiliation with technical journals has included positions as associate editor of the *Journal of Fluid Mechanics* and the *Quarterly of Applied Mathematics.* Dr. Carrier has also been a member of the *SIAM Journal of Applied Mathematics* and "Mathematics in Science" (Springer-Verlag) editorial boards.

His society memberships include being a Fellow in the American Academy of Arts and Sciences and belonging to the Society of Industrial and Applied Mathematics, American Society of Mechanical Engineers, National Academy of Sciences and Sigma Xi. He received the Pi Tau Sigma Richards Memorial Award (ASME) in November 1963.

At Cornell University, Dr. Carrier received his M.E. in 1939 and Ph.D. in 1944.

5

tsunamis
and wave run-up

Introduction

When a large displacement of the ocean floor occurs, an energetic gravity wave is generated which can travel several thousands of miles and cause severe damage to land areas on which it impinges. The severity of the run-up depends on the gross features of the initiating ground motion; the size, slope and shape of the land area where the response is of interest; the average depth of the ocean; and the travel distance. It may also depend on the topography of the intervening ocean floor. In order to obtain a quantitative understanding of the phenomenon, one must take into account the dispersive character of the propagation over much of the trajectory; the nonlinear effects which are important during the run-up phase; the refraction of the wave associated with the general "target" geometry; and possibly the cumulative reflection and scattering by the irregular bottom topography.

It would be exceptionally difficult to account for all these contributions to the phenomenon using a single comprehensive theory, but one can understand the implications of many of the foregoing mechanisms by treating each of the different facets of the wave problem by individual simpler theories.

Thus, the following will be studied in sequence:

1. The wave which is propagated over an indefinitely extended basin of uniform depth when an arbitrary ground motion is prescribed. This study will be conducted in only one-horizontal space dimension noting that the results for the more realistic (2-horizontal coordinate) problem are more difficult to obtain directly but can be inferred, using the fact that the far field wave pattern differs little from the one-dimensional pattern except that an extra attenuation is intensity proportional to $r^{-\frac{1}{2}}$ is present in the two-dimensional case.

2. The run-up on a plane shelf with slope α of the wave which, according to the theory under (1), propagates to the target area.

3. The effects of bottom topography on a one-dimensional monochromatic wave and the implications of such effects on the wave studied in (1).

4. The effects of systematic features of the bottom topography.

5. The intensification pattern associated with a linear analysis of wave refraction by a conical island.

6. The piecing together of (1) through (5) to see to what extent these analyses can account for the observations.

In these studies much mathematical detail will be suppressed, since it is already recorded elsewhere. Appropriate references are given.

Propagation Over Deep Water

Considered here is the wave which arises when, as in Figure 5-1, the bottom of the basin moves with a vertical component of velocity $V(x,t)$. Since dispersion is important in this phase of the wave motion and since the waves of interest may have amplitudes of a few feet and a lateral length scale of about 100 miles, the slope η_x is so small that a linear theory is entirely adequate. (Actually, all criteria for the suitability of the linear theory are satisfied for the problem of this section.) Furthermore, frictional losses can be shown to be negligible during this phase.

Accordingly, the theory which requires

$$\triangle \phi = 0 \text{ in } -\infty < x < \infty, \ 0 > y > -1, \ t \geq 0$$

$$\phi_y (x, -1, t) = V(x, t) \qquad (5\text{-}1)$$

$$\phi_y (x, o, t) + \phi_{tt} (x, o, t) = 0$$

is adopted.

Implicit in the use of these equations is the use of the basin depth b as the unit of length and of $\sqrt{b/g}$ as the unit of time.

The most straightforward approach to this problem is through the use of integral transforms. Restricting $V(x,t)$ and $V_t(x,t)$ to be zero for $t \leq 0$ defines

$$\bar{\phi}(\xi, y, s) = \int_{-\infty}^{\infty} e^{-i\xi x} \, dx \int_{o}^{\infty} e^{-st} \phi (x, y, t) \, dt \qquad (5\text{-}2)$$

and the equation for ϕ (the Fourier transform with respect to x and Laplace transform with respect to t of ϕ) becomes

$$\bar{\phi}_{yy} - \xi^2 \bar{\phi} = 0 \qquad (5\text{-}3)$$

The boundary conditions take on simple forms too, namely

$$\bar{\phi}_y (\xi, o, s) + s^2 \phi (\xi, o, s) = 0 \qquad (5\text{-}4)$$

and

$$\bar{\phi}_y (\xi, -1, s) = \bar{V} (\xi, s) \qquad (5\text{-}5)$$

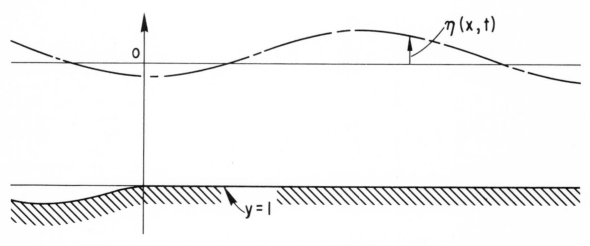

Figure 5-1. Geometry of the wave study of section 2. Dotted line denotes displacement of the bottom at time t associated with the bottom velocity V(x, y). Dashed line denotes the wave surface displacement associated with the wave. Both curves are only schematic.

and conventional manipulations lead to

$$\bar{\eta}\,(\xi,\,s) = \frac{-s\,\bar{V}(\xi,\,s)}{\xi\,(s^2\,\cosh\,\xi\,+\,\xi\,\sinh\,\xi)}$$

$$(5\text{-}6)$$

In order to find a more useful description of $\eta(x,t)$, the inversion integrals of transform theory must be evaluated. This can be done explicitly only if $\bar{V}\,(\xi,s)$ is known. It suffices at present to choose

$$V\,(x,\,t)\;=\;\begin{cases}X\,(x)\;c^2 t\,e^{-ct}\,,\;t\,>\,0\\0\qquad\qquad,\;t\,<\,0\end{cases}$$

$$(5\text{-}7)$$

so that the time scale for the ground motion is c^{-1} and, since

$$\int_o^\infty c^2\,t\,e^{-ct}\,dt\;=\;1$$

the intensity and extent of the ground motion are still contained in a general way in $X(x)$. With this choice for V, inversion of $\bar{\eta}$ over s gives

$$\eta(x,t) = \frac{1}{2\pi}\int_{-\infty}^{\infty}e^{i\xi x}\,\frac{\cos\,[t\,f(\xi)]}{\cosh\,\xi}\,\bar{X}(\xi)\frac{c^2}{[c\text{-}if(\xi)]^2}\,d\xi$$

$$(5\text{-}8)$$

where

$$f\,(\xi)\;=\;(\xi\,\tanh\,\xi)^{1/2}$$

$$(5\text{-}9)$$

To evaluate this integral meticulously is out of the question, but, for large x and t, the method of stationary phase can be used. To find the wave height and wave form near the front of the wave is beyond the scope of the conventional stationary phase method, but an extension of that method (see Carrier[1] for the method and the details for this problem) one obtains the asymptomatically valid result

$$\eta(x,t) \sim \left[\frac{f(Z)\text{-}Zf'(Z)}{-3\,f''(Z)}\right]^{1/2}\frac{c^2\,\bar{X}(Z)}{[C\text{-}if(Z)]^2}\,J\!\left(X_o\!\left\{\frac{f(Z)}{f'(Z)}\text{-}Z\right\}\right)$$

$$(5\text{-}10)$$

where Z is the positive root of $f'\,(Z) = X_o/t$. The leading part of the wave lies near $x = t$; and, for $x \simeq t$, Z lies close to the value zero. For such values of Z,

$$\frac{C^2\,\bar{X}\,(Z)}{[C\text{-}if(Z)]^2}\;\simeq\;\int_{-\infty}^{\infty}X(x)\,dx\;=\;\int_{-\infty}^{\infty}dx\int_{-\infty}^{\infty}V(x,t)\,dt$$

$$(5\text{-}11)$$

so that, ordinarily, the only feature of the initial motion which matters in initiating the wave is the volume displacement of bottom material. This is *not* the case, of course, when some of the bottom displacement is upward and some down with the foregoing integral ending up small. For such cases, more details of the behavior in x of $V\,(x,t)$ must be used in obtaining the expression for $\eta(x,t)$. In general, however,

$$\eta(x,t) \sim \left[\frac{f(Z)\text{-}Zf'(Z)}{-3\,f''(Z)}\right]^{1/2}J\!\left(X_o\!\left\{\frac{f(Z)}{f'(Z)}\text{-}Z\right\}\right)$$

$$(5\text{-}12)$$

can be taken for a unit volume displacement of material. The maximum value of $\eta(x_o,t)$ under these circumstances (for a downward ground motion) is

$$\eta\,\max\;\sim\;.28\;X_o^{-1/3}$$

$$(5\text{-}13)$$

Effects of Bottom Topography

If, in section 2, a theory which ignored dispersion had been used, Equation (5-8) would contain the factor $\cos\,(t\xi)$ instead of $\cos\,[tf(\xi)]$. This would have an overwhelmingly important effect on the result, Equation (5-10), implying that the consequences of dispersion cannot be ignored. Because of that dispersion, as the applicability of the method of stationary phase implies, the wave height at a given x and t depends only on a very limited wave number (or frequency) range which is centered on $\xi = Z$, where Z is defined after Equation (5-10).

Considered now is the propagation of the wave generated by the same ground motion as that of

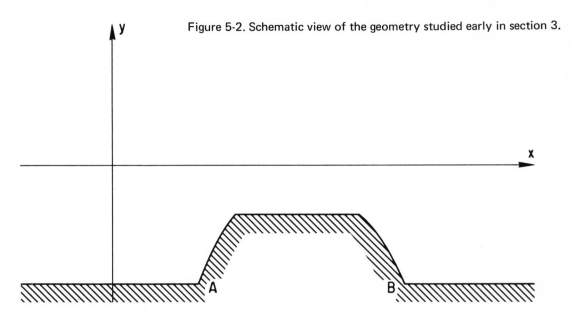

Figure 5-2. Schematic view of the geometry studied early in section 3.

section 1 but with the geometry of Figure 5-2. The same theory can be used to describe the wave which arrives at A, the reflection and transmission of the wave can be analyzed by the "bump" in AB and the theory of (2) again can be used substantially to analyze the wave which continues past B. When this is done, the second step (analysis of reflection and transmission at AB) will provide a result which implies that the wave height at B [called $N(B,t)$] will be given by

$$N\ (B,t)\ =\ \eta\ (B,t)\ T\ (Z_b) \qquad (5\text{-}14)$$

where $\eta(B,t)$ is defined by Equation (5-10), $T(k)$ is the transmission coefficient of the bump in AB for T waves of wave-number k and the value of Z_b to be used in Equation (5-14) is that derived from

$$f'\ (Z_b)\ =\ B/t$$

Equation (5-14) is valid whenever the dependence of $T(k)$ on k varies slowly enough in k so that the dominant frequency band as determined by the stationary phase procedure is still dominant despite the presence of the bump. When this *is* the

case, it also follows that, at points x to the right of B, the wave height is

$$N\ (x,t)\ =\ \eta\ (x,t)\ T\ (Z) \qquad (5\text{-}15)$$

where Z is defined after Equation (5-10).

Dispersion is not important over the relatively short distance AB, and a simpler nondispersive theory can be used to calculate the transmission coefficient $T(k)$. Such calculations have been carried out by Kajiura[2] for a variety of bump configurations. The essential feature of the result is that only bumps which are narrower than $2\pi/k$ and whose heights are substantial fractions of the water depth can provide nonnegligible reflections. Thus, it is concluded that the only individual topographical features of the Pacific which are important in the propagation of tsunamis are those associated with island or continental shelves. Since these particular items will be treated in later sections, the foregoing analysis provides no modification of Equation (5-10).

Despite the foregoing, it is still possible that cumulative effects occur because of the topographical irregularities of small vertical scale which extend over the whole basin. One can argue quickly that no important effects occur when such features have a lateral scale which is very large compared to the wave lengths of interest. One sees this simply because a WKB type of analysis of such propagation (be it gravity wave, acoustic wave, or electromagnetic wave) indicates clearly that such reflections are small. Carrier[1] verifies and documents this for the gravity wave problem, but the analysis used is not capable of accounting for the topographical features of lateral scale comparable to the wave length of the important part of the spectrum.

It is also relatively easy to argue that this aspect of the topography *might* be important. If, in fact, the ocean floor were "sinusoidal" so that the depth is described by

$$h = 1 + \cos \alpha x' \qquad (5\text{-}16)$$

shallow water theory would require that, to a high degree of approximation, the wave height must obey the equation

$$(h \eta_x)_x - \eta_{tt} = 0 \qquad (5\text{-}17)$$

so that, with $x' = \int_o^x dx/h$

$$\eta_{x'x'} - h \eta_{tt} = 0 \qquad (5\text{-}18)$$

It is noted that the depth is sinusoidal in the slightly distorted variable x' rather than in x. This is a convenient choice and sacrifices nothing, since no real ocean floor is sinusoidal in either coordinate.

The appropriate solution of Equation (5-18) for waves in which the time dependence is e^{it} (time and length scales have been chosen to suppress irrelevent parameters) is not particularly interesting (i.e., there is negligible reflection) unless $\alpha \simeq 2$. When α is precisely 2, η decays in the propagation direction because of cumulative reflection according to the recipe

$$\eta \sim e^{-\epsilon x/2} \qquad (5\text{-}19)$$

If the average depth is 4 miles, the undulation height is 400 ft., the travel distance is 3,000 miles and then $\epsilon = 1/50$ and x_{max} is 750. Equation

(5-19) predicts that most of the wave at and near this frequency would not reach the target.

To make a more realistic calculation requires a more realistic description of the topography and a technique which can cope with the increased complication. A useful compromise emerges when a statistical description of the topography is adopted, it is written, in fact,

$$h = 1 + \epsilon f(x) \qquad (5\text{-}20)$$

where ϵ characterizes the average height of the deviations from the average depth and where what is known about f is only its auto-correlation function

$$R(y) = <f(x) \, f(x+y)> \qquad (5\text{-}21)$$

In this recipe, $<>$ denotes the ensemble average.

The use of shallow water theory again leads to Equation (5-18) with h defined by Equation (5-20), and a careful analysis which Carrier[3] has carried out shows that $< \eta^2 (x') >$ decays according to

$$<\eta^2 (x')> \sim e^{-\epsilon^2 x^{1/2}} \qquad (5\text{-}22)$$

provided that $R(y) << 1$ when $y \geqslant \epsilon^{1/2}$. The implications of Equation (5-22) are quite different from those of Equation (5-19). The extra ϵ reduces the attenuation to a few percent and, when it is remembered that the non-one-dimensional nature of the real topography reduces the coherence of the structure still further, it is not possible to claim an important effect from this source. The restriction on R which follows Equation (5-22) merely reflects the fact that the exponent in Equation (5-22) could be bigger if the bottom undulations were almost sinusoidal.

There is one more topographical consideration which could be of interest. If—accompanying the random zero-average features of the topography—there were also a systematic feature in the form of a low ridge which ran from the area of seismic activity to one of the target areas, that ridge would act as a wave guide, and the attenuation along the ridge could be much less than that going in other directions. No satisfactory analysis of the transient problem with such a ridge has been achieved, but efforts at such an analysis have made it quite clear that—to be significant in this way—the ridge would have to be much higher than anything one can infer from existing maps of the Pacific.

It is concluded that, as interesting and as *potentially* important as they may have been, the topographical features of the deep ocean probably provide no really important modification of the result given by the constant depth propagation theory.

Run-up on a Plane Beach

When the wave of section 2 reaches a plane sloping shelf and propagates into increasingly shallow water, the lateral scale shortens; the wave amplitude gets higher; and, in the very shallow regions, linear theories become inadequate. For the part of the spectrum of interest, however, the distance traversed on the shelf is too short for dispersion to have important consequences. Furthermore, a typical boundary layer thickness is of the order of a few centimeters. The dissipation which can occur in so thin a layer is so small compared to the total energy content that it is hard to imagine that friction is important. (That question will be considered later.) Therefore, the non-frictional, nonlinear, nondispersive but appropriate shallow water theory can be used.

That theory (see Stoker[4]) has the form

$$[(b + \eta)\, u]_x + \eta_t = 0 \qquad (5\text{-}23)$$

$$u_t + u u_x + \eta_x = 0 \qquad (5\text{-}24)$$

where $b(x)$ is the depth of the water, $u(x,t)$ is the horizontal component of particle velocity (which is independent of depth in this theory) and η is the wave height.

When $b = (x_1\text{-}x)\,\vartheta$, as in Figure 5-3, it is a surprising but gratifying fact that solutions of these equations can be constructed in the following way (see Carrier[1,5] and Greenspan[5]):

One can find any function $\psi\,(\sigma,\lambda)$ which satisfies

$$(\sigma \psi_\sigma)_\sigma - \sigma \psi_{\lambda\lambda} = 0 \qquad (5\text{-}25)$$

and define (see Figure 5-3):

$$u = \psi_\sigma \big/ \sigma \sqrt{\vartheta} \qquad (5\text{-}26)$$

$$\eta = \tfrac{1}{4}\,\psi_\lambda - u^2/2 \qquad (5\text{-}27)$$

$$x - x_1 = \psi_\lambda \big/ 4\vartheta - \sigma^2/16 - u^2/2\vartheta \qquad (5\text{-}28)$$

$$t = \lambda/2\sqrt{\vartheta} - u/\vartheta \qquad (5\text{-}29)$$

The functions u,η as functions of x,t—implicitly described by the foregoing recipes—are solutions of Equations (5-23) and (5-24) provided only that the quantity (the Jacobian) $J = x_\eta\, t_\lambda - x_\lambda\, t_\eta$ neither vanishes nor becomes infinite anywhere over the region $x_o - x_1 < x - x_1 < 0$, $t > 0$.

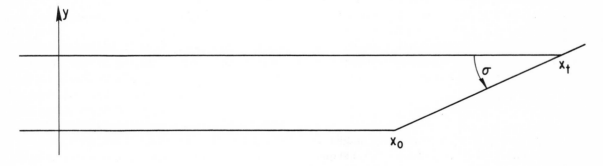

Figure 5-3. Geometry for run-up study.

The reader can verify by direct substitution that the function ψ for which

$$\frac{1}{4}\,\psi_\lambda = \frac{1}{2\pi} \int_{-\infty}^{\infty} \frac{\cos \xi\, x_0}{\cosh \xi}\; X(\xi)\; \frac{e^{-i\lambda f(\xi)/2\sqrt{\vartheta}}\, 2J_0\left(\sigma\, f(\xi)/2\sqrt{\vartheta}\right)}{H_0^{(2)}\left[Lf(\xi)/2\sqrt{\vartheta}\right]}\; \vartheta\,\xi \tag{5-30}$$

is a solution of Equation (5-25). He can also verify (more laboriously) that sufficiently small multiples of this do not violate the requirement on the Jacobian stated above. Furthermore, the linear theory is valid at $x = x_0$ (where σ has the value L). Therefore, to a high degree of approximation, $x - x_1$ and t can be taken as

$$x - x_1 \simeq -\sigma^2/16 \tag{5-31}$$

$$t \simeq \lambda/2\sqrt{\vartheta} \tag{5-32}$$

With this identification, ψ_λ—as given by Equation (5-30)—has an incoming (toward x_1) part which at L is the same as that given by Euqation (5-8) of section 2. That is, displayed in Equation (5-30) is the solution of the shallow water equations which connects "properly" to the wave which has arrived from the seismic source. An even more "proper" treatment would take account of the reflection which occurs because of the sudden change of slope at x_0. The result of that would be a replacement of $H_0^{(2)}$ $(Lf/2\sqrt{\vartheta})$ by

$$J_0\,(Lf/2\sqrt{\vartheta}) + i\,J_0'\,(Lf/2\sqrt{\vartheta})$$

in the denominator of Equation (5-30), a replacement which affects the result in no appreciable way.

The integral of Equation (5-30) can also be evaluated for large t by the method of stationary phase. Furthermore, since $\sigma = 0$ always corresponds to the actual (not the ambient) water line and since the velocity u is zero whenever η has a local maximum or minimum, Equation (5-30)—evaluated at $\sigma = 0$ and at those times— t_j, where η is a maximum, is a recipe for

$\eta(t_j)$. The run-up for the same initiating motion considered in section 2 comes out to be

$$\eta_{max} = \sqrt{2/x_0\,\vartheta} \tag{5-33}$$

and, in particular, the ratio of the greatest run-up to the largest deep water wave amplitude at $x = x_0$ is

$$\frac{\text{max run-up}}{\text{max } \eta(x_0,t)} = 5.6\,\vartheta^{-\frac{1}{2}}\,X_0^{-1/6} \tag{5-34}$$

Refraction by Islands

The phase velocity of a long gravity wave is approximately equal to $\sqrt{gh}$, where h is the local depth. When a wave impinges on a shelf whose level lines are curved, the rays associated with the waves focus toward the shallower regions. This refractive behavior implies that the amplification associated with propagation up the shelf is not given accurately by the theory of section 4, but that this amplification will be enhanced on some parts of the island perimeter and diminished on others.

When one looks at the details of the sloping beach theory of section 4 (see Carrier and Greenspan[5]), it is clear that nonlinear effects become important only in very shallow water. Thus, if one calculates the refractive influence by a linear theory and compares the run-up so predicted with that predicted for the plane shelf, the ratio of these should provide an excellent estimate of the intensification factor due to refraction.

Guided by the foregoing argument and by the justification put forth in section 3 for confining

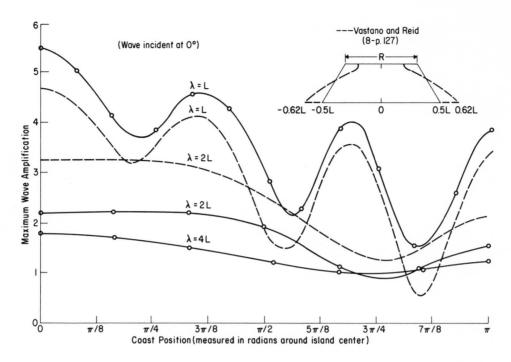

Figure 5-4. Maximum wave amplification at coast (Hawaii) (after Lautenbacher[6]). λ is wavelength; L is depicted; R is depicted.

one's attention to monochromatic waves, Lautenbacher[6,7] has calculated the refractive influence, using a numerical treatment of an integral equation derived from the conventional linear shallow water theory. More explicitly, he treats the refraction of an incident, monochromatic wave on an ocean of constant depth by an isolated island whose level lines are ellipses.

The integral equation is so chosen that the domain is finite and the radiation condition is automatically incorporated into it. Furthermore, the formulation permits the treatment of geometries in which the depth goes continuously to zero at the perimeter of the island. The equation he works with is illustrated in the box at the bottom of this page

$$w = \frac{-i}{4h'} \iint \left\{ -k^2 h H_0^{(1)}(kR) + k h_{r'} \frac{H^{(1)}(kR)}{R} \left[r\cos(\vartheta-\vartheta') - r' \right] \quad r'w dr' d\vartheta' \right.$$

$$+ \frac{k}{4h'} \iint H_0^{(1)}(kR) e^{ik(r'\cos\vartheta + \frac{1}{2})} (\cos\vartheta' \, h_{r'} + ikh) \, r' dr' d\vartheta'$$

(5-35)

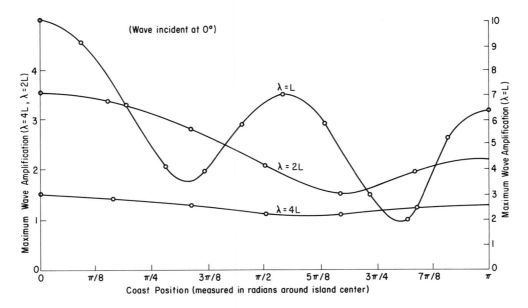

Figure 5-5. Maximum wave amplification at (Oahu) coast (after Lautenbacher[6]).

where

$$w(r, \vartheta) = \eta - \eta_{inc}, h'(r, \vartheta)$$

is the depth, $h = 1 - h'$, and

$$R = [r^2 + r'^2 - 2rr' \cos(\vartheta - \vartheta')]^{1/2}.$$

In the integrand, h and w are evaluated at r', ϑ'; outside of the integrand, they are evaluated at r, ϑ. The domain of integration includes *only* the region between the deep water and the shoreline. Lautenbacher thoroughly describes the numerical treatment, and the results of its use are depicted in Figures 5-4, 5-5, 5-6.

Vastano and Reid[8] use an alternative approach. They integrate the differential equation directly by numerical means. Their procedure *cannot* be used for geometries in which the depth goes continuously to zero (whereas Lautenbacher's procedure is confined to such geometries) but

would be viable for islands whose perimeters are cliffs adjacent to which the water is shallow.

The Lautenbacher results are more appropriate for the comparison wanted here. Thus, the run-up ratio at a given place on the perimeter of a given island is calculated by multiplying the ratio given in Equation (5-34) by the appropriate amplification factors from one of Figures 5-4, 5-5, 5-6.

Tsunamis

It is difficult to interpret the foregoing in the light of the observations. To illustrate this, it is noted that it has sometimes been suggested that observations made at Wake Island by Van Dorn[9] could be identified with the wave height at x_o and that run-up observations on the Hawaiian Islands (Figure 5-7) could be compared to this "deep water" result to check the ratios implied by the foregoing theory. When this is done, the ratio of the run-up to the deep water amplitude of the theory is too small (by a factor of 2 or more) to

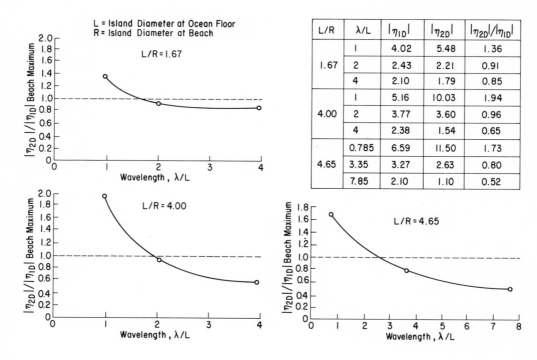

L/R	λ/L	$|\eta_{1D}|$	$|\eta_{2D}|$	$|\eta_{2D}|/|\eta_{1D}|$
	1	4.02	5.48	1.36
1.67	2	2.43	2.21	0.91
	4	2.10	1.79	0.85
	1	5.16	10.03	1.94
4.00	2	3.77	3.60	0.96
	4	2.38	1.54	0.65
	0.785	6.59	11.50	1.73
4.65	3.35	3.27	2.63	0.80
	7.85	2.10	1.10	0.52

Figure 5-6. Ration of two-dimensional to one-dimensional maximum wave amplitude on beach (after Lautenbacher[6]). L and R defined on Figure 5-4.

account for the observations; and this could be due to any or all of several hypothetical possibilities, namely:

1. The deep water interpretation of the Wake measurements may be inappropriate.
2. The observed high run-ups may be due to the amplification associated with the local geometries of the large amplitude areas.
3. The deep water wave which arrives at Wake may be weaker than that coming to Hawaii because of wave guide effects.
4. Tremors could have a geometric distribution and temporal history, which favor constructive interference at Hawaii and destructive interference at Wake.
5. Observations at Hawaii are recorded primarily when they are impressive. Very small run-ups are lost (or at least undistinguished) in the general noisy background provided by

the usual oceanic activity. Thus, the Hawaiian observations may have a strong bias insofar as average run-up is concerned.

It is also stated frequently that "the run-up associated with the first wave is not the largest; more often it is the third." If this is true, there again are various possibilities.

6. The ground motion may consist of a sequence of at least two distinct tremors so spaced in time that the head wave generated by one tremor augments (e.g., the third crest of a previous tremor). That is, the temporal spectrum of the ground motion may be such that a more intricate treatment is needed for the integration of Equation (5-8).
7. The run-down following the first crest may be sufficiently impeded by friction that the

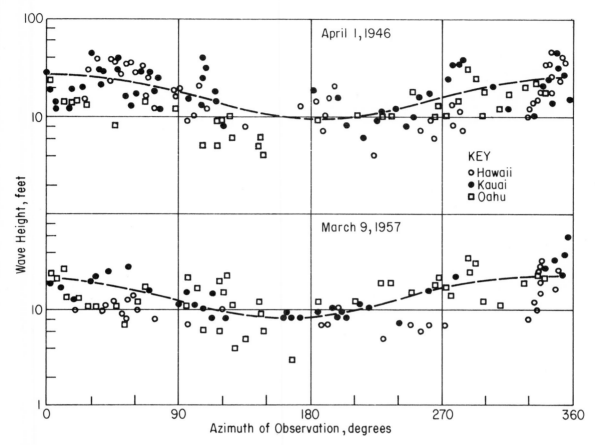

Figure 5-7. Azimuthal distribution of 254 run-up observations in Hawaiian Islands for two large Aleutain tsunamis (after Van Dorn[9]). The 0° azimuth corresponds to incident wave direction from the respective epicenters.

ambient water line for the next crest arrival provides for a higher run-up. This is the "set-up" phenomenon which occurs in the steady state response of shelf areas to incident waves.

8. The local response of "resonant" local geometries to a transient can require several "periods" to have passed before the response gets close to the steady state response. The response in such areas could, and probably does, dominate the observational evidence.

Each of (1) to (8) can be discussed, but no accurate assessment of the importance of each

exists. This writer has tried to speak up to (3) in section 3; (4) and (6) would be tenable in any single instance, but that either should *consistently* occur over many tsunamis seems to be a low probability coincidence. The writer finds it hard to attach much importance to (7), and—without enough evidence to carry much conviction—he favors items (1), (5) and (8) with the occasional relevance of (2) and (6).

It is clear that there is still an inadequate understanding of Tsunamis. When the inadequacies have been removed, however, the foregoing considerations should prove to be significant items in that understanding.

References

1. Carrier, G. F. "Gravity Waves on Water of Variable Depth," *J. Fluid Mech.,* 24, No. 4 (1966), 641-659.

2. Kajiura, K. "On the Partial Reflection of Water Waves Passing Over a Bottom of Variable Depth," *IUGG Monograph,* 24.

3. Carrier, G. F. "Some Stochastically Controlled Dynamical Systems," to appear in the *J. F. M.*

4. Stoker, J. J. "The Formation of Breakers and Waves," *Comm. Pure Appl. Math,* 1 (1948), 1.

5. Carrier, G. F. and H. P. Greenspan. "Water Waves of Finite Amplitude on a Sloping Beach," *J. Fluid Mech.,* 4 (1958), 97.

6. Lautenbacher, C. "Gravity Wave Refraction by Islands," to appear in *J. Fluid Mech.*

7. ———. "Gravity Wave Refraction by Islands," thesis presented to Harvard University, 1966.

8. Vastano, A. C. and R. O. Reid. "Tsunami Responce for Islands," *J. Marine Research,* (1967), 129-139.

9. Van Dorn, W. G. "Tsunamis," *Advance in Hydroscience,* 2 (1965), 1-48.

6

hurricanes

Introduction

In this chapter, the author will present hypotheses (first given by Carrer[1]) concerning (1) the dynamical structure of a mature hurricane and (2) the necessity for significant heat and mass transfer from the ocean to the boundary layer air in such a storm.

A typical mature storm contains swirling air whose maximum speed is 100 to 200 miles an hour;, its radius (the region over which the dynamics are immediately relevant) is 500 miles or more. It is generally agreed that a hurricane has a core, whose radius is somewhat more than 10 miles, in which the air is relatively warm, dry and motionless. The structure proposed here is shown in Figure 6-1. Region I is the core; II is the updraft region. The air flowing through the updraft region is fed in through the boundary layer which occupies region III and has an angular momentum content which depends on the boundary layer details. In region IV the air is swirling rapidly and sinking very slowly. The angular momentum content is there because—during the generation phase of the storm—air moved from a large distance r_0 to a smaller radius r, conserving much of its absolute angular momentum (not momentum relative to the earth) as it moved.

Dynamics

In order that the foregoing picture be self-consistent dynamically, the laws governing conserva-tion of mass, momentum and energy must be satisfied; and the state of the gas in each part of the configuration must be consistent with the dynamical configuration. In order to use these laws, the region worked with is depicted by the dotted lines of Figure 6-1, and it is pretended that these lines denote rigid walls. This is valid unless there is some dynamically significant interaction with the rest of the atmosphere above the top of the box in Figure 6-1 (which can be put at any altitude—e.g., the 150-millibar level). It is hard to assess the likelihood of such an interaction; but, in any event, this study can show whether in this situation (which is analogous to the hurricane environment) a swirling pattern like Figure 6-1 can be self-sustaining.

Figure 6-2 denotes (curve a) the pressure-temperature relation in a typical low latitude (15°) atmosphere in the hurricane season. It also shows (curve b) the moist adiabat (the pressure-temperature history a particle of air with typical moisture content will experience if it is adiabatically expanded from state A to state B if this process moisture condenses at a rate which maintains saturation). This curve describes the set of thermodynamic states which the updraft air will experience as it climbs. Finally, Figure 6-2 also shows (curve c) the dry adiabat which air that has risen to the top of the container will follow when it descends again, compressing adiabatically as it does so. No water is evaporated on this path and that is why (c) differs from (b).

It is now hypothesized that the air in the core is merely air which has come through the updraft region earlier in the storm and recirculated slowly into the core. Thus, if the pressure at the top is P_o, the hydrostatic equation

$$Py + \rho g = 0 \qquad (6\text{-}1)$$

and the thermodynamic relation given by (c) of Figure 6-2 allows the pressure to be calculated throughout region I and, particularly, at y = 0.

Clearly if Figure 6-1 is to be self-consistent, the pressure in the core at r = 0 must differ from that at R (where the column of air is also almost motionless) by an amount which can support the centripetal acceleration of the gas between R and 0 at the upper edge of the boundary layer. To use this fact more quantitatively, one must adopt a dependence of V (the peripheral speed) on r. For this purpose only, $V = V_{max} (r/r_1)^{-d}$ is used where, guided by observation in the laboratory and the field, $1/2 < \alpha < 1$ is expected. The greatest V_{max} which can be sustained is obtained when $\alpha = 1$.

Fendell and Dengarahedian[2] have used this model to estimate the speed achieved in several storms with considerable apparent success. (They also use another observationally based method to independently check such estimates.) Most of these examples are tornadoes which require a modification of the foregoing (which they use), but corresponding estimates for hurricanes (which have not been published yet) also confirm the validity of this isolated aspect of this model.

The question which needs attention is this: For what ambient atmosphere (at radius R) and swirl velocities in region IV is there a dynamically consistent boundary layer in III *which accepts a moderately uniform downdraft from IV* and spews air into II with an angular momentum content which leads to interfaces having qualitatively the location shown in Figure 6-1?

In such flows, of course, the interfaces will move slowly as the air in IV is depleted and II becomes more voluminous. The crucial part of the question is the clause "which accepts a moderately uniform downdraft from IV." The analysis of this boundary layer problem is difficult and is still being pursued, but the author has little doubt that the analysis will be forthcoming and that the swirl distributions will exist and will be identified.

The second question to which this chapter is addressed concerns heat and mass transfer from the ocean to the air. Specifically, one asks whether

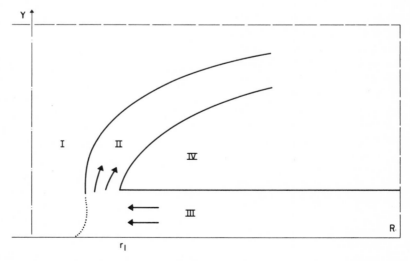

Figure 6-1. Supposed geometry of hurricane. I—still air, II—draft, III—boundary layer, IV—swirling "reservoir."

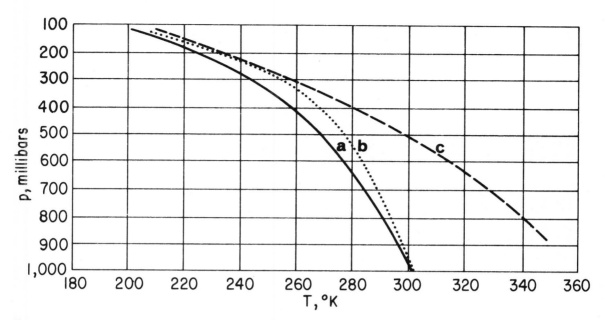

Figure 6-2. Typical pressure-temperature plots for West Indian hurricanes. Indicated are (a) typical ambient profile, (b) moist adiabat based on ground-level ambient conditions and (c) dry-adiabatic recompression (without condensation) of ambient air at the "top" of the ambient profile.

a significant flow of vapor and sensible heat *can* flow across the air-sea interface and whether anything more in the way of such flow is needed than the heat-and-vapor flow, which maintains the ambient (large radius) state of the atmosphere when no storm is present.

Allan Hammond has made a sequence of analyses (each more realistic than its predecessor) of the flow of heat and moisture to the air and finds that under typical conditions less than (and probably much less than) 10% of the pressure discrepancy [P(R) − P(O)] can be attributed to the boundary layer transfer. Furthermore, the fact that speeds that are easily as large as the speeds in actual storms are estimated without invoking any such transfer is ample evidence that storms can proceed with a transfer in the outer part of the storm. Such a transfer is merely that which would occur even in the absence of the storm.

Thus, unless and until it is found that the expected boundary layer structure in III is not really there, this simple but as yet unsubstantiated picture of a hurricane may contain all the principal features and mechanisms of the phenomenon.

References

1. Carrier, G. F. "Severe storms—A Preliminary and Informal Report," a report to TRW Systems, Inc.

2. Fendell, F. and P. Dengarahedian. "On the Structure of Severe Storms," a preprinted presentation of TRW at the AIAA meeting, June 1969.

7

the midlatitude ocean circulation

Introduction

It is generally believed that the midlatitude circulations in the North Atlantic and North Pacific oceans are wind driven, and a sequence of theories has been put forth over the years in attempts to explain these circulations. Nevertheless, the state-of-understanding of these flows is far from satisfactory.

In this exposition the ideas and consequences of several of these theories will be examined and contrasted. In particular, the nature of the difficulties which continue to impede progress will be discussed.

The regions of interest are characterized in Figure 7-1, and the yearly average wind stress, which is believed to be exerted on the ocean surface, is also depicted there. (Any "observational facts" cited in this chapter must be interpreted as the best compromise available to the author which

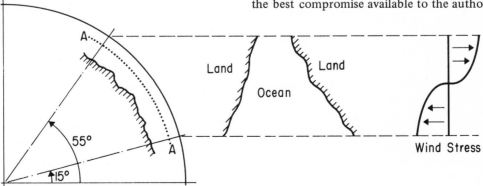

Figure 7-1. Region in which midlatitude circulation takes place. Z axis is parallel to earth's rotation vector Ω. A-A denotes thermocline position.

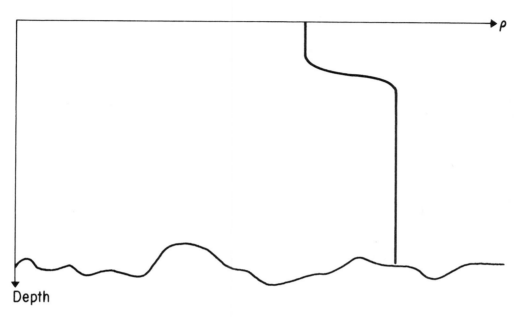

Figure 7-2. Density structures as function of depth.

incorporated the gross features of the observations and the descriptive simplicity so necessary for readily interpreted theory.) For reasons which will appear later, it is reasonable to expect no net transport of fluid across the constant latitude boundaries. Thus, the basins bounded by these boundaries and by the continental slopes are "closed."

The density distribution of the water is typically that of Figure 7-2. It is of particular interest that there are two layers of relatively uniform density separated by the thermocline, a thin region in which the density gradient is much larger than it is in the two layers.

Any dynamical theory of the circulation must use the fundamental laws implying the conservation of mass and momentum, and it must include a quantitative description of the frictional stresses as functions of the velocity gradient and other relevant variables. At this stage of development, the very complicated processes which account for the stress distribution and for the transfer of heat and salt are certainly not understood. It is advantageous to approximate them by using the conventional Newtonian (laminar) relationships augmented by empirically estimated diffusivities for momentum transfer, salt transfer and heat transfer. Such a system is adopted here, noting only that the extent to which such a formalism is adequate has by no means been determined.

Superimposed on the average flow to be studied is an extensive collection of transient phenomena. The contributions made by these to the circulation system in this study will be ignored in all theories to be discussed (except for the foregoing turbulent diffusion modeling), but the reader should note again that no one has really demonstrated that the time-averaged consequences of these transients is not an important part of the "steady state" phenomenon.

With these preliminaries, one can write the equations for the conservation of mass and momentum for the incompressible fluid in the form

$$\text{div } \vec{v} = 0 \tag{7-1}$$

$$\rho \vec{v} \cdot \text{grad } \vec{v} + 2\rho \vec{\Omega} \times \vec{v} + \text{grad } p +$$
$$\rho \vec{\Omega} \times \vec{\Omega} \times \vec{r} + \rho \vec{g} = \mu \Delta \vec{v} \tag{7-2}$$

where $\vec{v}(\underline{r})$ is the particle velocity relative to the rotating system whose angular velocity is $\Omega = 2\pi$ rad/day, μ is the "viscosity," ∇ is the laplace operator and $\vec{g}$ is a vector whose magnitude is the acceleration of gravity and which is directed radially outward from the "origin" of Figure 7-1.

The transfer process for density can usefully be approximated by

$$k \Delta \rho + \vec{v} \cdot \text{grad } \rho = 0 \qquad (7\text{-}3)$$

where k is the diffusivity which mimics compositely all small scale mixing mechanisms not accounted for in the term $\vec{v}$ grad ρ.

Rotating Fluids

Before discussing the ocean current problem per se, it will be helpful to review some general facts about motions in rotating fluids.

It has long been known[1] that, in regions where friction and nonlinear acceleration processes play insignificant roles, the velocity field in a homogeneous fluid obeys the "Taylor-Proudman theorem"

$$\partial \vec{v} / \partial z = 0$$

where z is the coordinate in the $\vec{\Omega}$ direction. When the diffusivity is small enough and the depth 2L of the container is large enough so that $\mu/\rho \Omega L^2 \ll 1$, it is also true[2] that flows driven by surface tractions have the following structure.

1. Near each of the top and bottom surfaces there is a thin frictional "Ekman layer," whose thickness is of the order of $\sqrt{v/\Omega}$ (where $v = \mu/\rho$). Between these layers, the velocity field is almost independent of z. The flow near the lateral boundaries depends on several factors, and its discussion is not profitable here.
The relevant container of constant depth in which the fluid is driven by a prescribed surface velocity [$\vec{V}_+$ (x,y) at the top $\vec{V}_-$ (x,y) at the bottom], the fluid velocity

everywhere except in the thin Ekman layers is a simple average of $\vec{V}_+$ and $\vec{V}_-$.

2. In a container whose depth varies with slopes large compared to $\sqrt{v/\Omega L^2}$, velocities comparable in magnitude to the surface velocity can occur only along contours of constant depth.

3. In a container containing two immiscible fluids of different densities, the fluid in each layer conforms to statements (1) and (2) as though the interface itself were a (moving) container wall.

When a fluid with continuously varying density is set into motion relative to its rotating container by surface tractions, gravitational attraction will prevent much distortion in the vertical of the constant density surfaces, provided only that the density gradient in the ambient system is large enough. When this is the case, the characterizations of the motion of a homogeneous fluid stated earlier in this chapter are not true. In fact, when the density gradient is very strong, the velocity of the fluid will vary continuously and smoothly from its prescribed value at the top to its prescribed value at the bottom. Carrier[3] delineates this in detail; in particular, he shows that, if $\vec{V}_+$ (x, y) is the velocity at the top, the velocity just under the upper Ekman layer is

$$\frac{V(x,y)}{V_+ (x,y)} = \frac{1+z}{2+z} \qquad (7\text{-}4)$$

where

$$z = g\beta L (v/\Omega)^{1/2} N \Big/ 3k\Omega R^2$$

Here, g is the acceleration of gravity $\beta = \rho_o, z/\rho_o$; L is the half depth of the layer; R is the lateral scale of the phenomenon; and N is the ratio of the lateral diffusivity k to the vertical diffusivity k.

Thus, when $z \ll 1$, $\frac{V}{V_+} \simeq \frac{1}{2}$, and the flow resembles that of a homogeneous fluid. Alternatively, when $z \gg 1$, there is a negligible Ekman layer, and stratification dominates.

Two 2-dimensional approaches

The conservation equations, the stratification pattern and the geometry associated with the oceanic circulation are far too complicated to permit a mathematical solution without further simplification. Furthermore, it is not the objective here to show that the motion is consistent with the laws of physics (that is not in doubt) but to determine which mechanisms and which facets of the geometry play the major roles in determining the nature of the phenomenon. Accordingly, it is useful to evolve a simple governing theory. This can be attempted in the following ways.

One can hypothesize that, to understand the horizontal transport of fluid, it suffices to invoke the horizontal components of momentum conservation, ignore the vertical component of velocity (which can be appreciable only in small lateral scales in view of the thinness of the layer) and look only at depth-averaged quantities. With this hypothesis, one writes the momentum equations in spherical coordinates; integrates them over the depth, ignoring the contribution of bottom stress;

introduces a potential $\vec{h}\,\psi$, the components of whose curl are

$$\int_{depth} U dr \text{ and } \int_{depth} V dr;$$

renders the equations dimensionless; and ignores terms which can be shown to be numerically small compared to retained terms.

This process is essentially that used in all of the papers in Robinson[4]. The equation obtained is

$$\epsilon \Delta \Delta \psi + \alpha(\psi_y \Delta \psi_x - \psi_x \Delta \psi_y) - \beta \psi_x = \sin y$$

$$(7\text{-}5)$$

The coordinate x is positive to the east; y is positive to the north; $\beta(y) = 2\Omega \cos\vartheta(y)$ and is a multiple of the rate of change of $2\vec{\Omega} \cdot \vec{r}$ with latitude (ϑ is latitude). α and ϵ are very small numbers, which depend, respectively, on the size of the wind stress and the eddy viscosity (for horizontal transfer). In the nonlinear term, one has taken for granted that it suffices to approximate $\int_{depth} u_i u_j dr$, (where $u_i \cdot u_j$, are horizontal velocity components) by

$$C \int_{depth} u_i dr \int_{depth} u_i dr$$

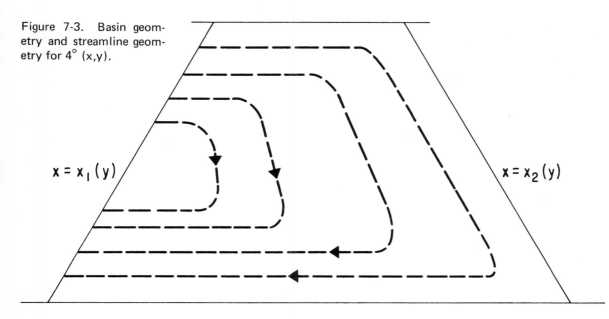

Figure 7-3. Basin geometry and streamline geometry for 4° (x,y).

$x = x_1(y)$

$x = x_2(y)$

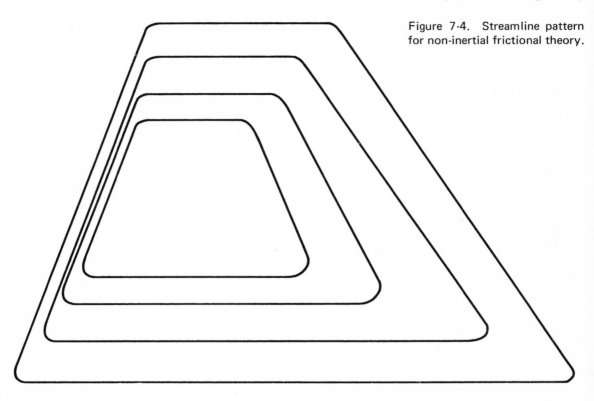

Figure 7-4. Streamline pattern for non-inertial frictional theory.

where C is a multiple of the depth of the domain in which the circulation occurs.

Since ϵ and α are small, one is tempted to study the flow which is predicted when they are ignored. The stream function for that flow is called $\psi^{(o)}(x, y)$, and it is noted that

$$\psi^{(o)}(x, y) = [A(y) - x] \sin y \qquad (7\text{-}6)$$

For the basin of Figure 7-3, one can choose A (y) so that $\psi^{(o)}(x, y)$ vanishes on the eastern boundary. When this is down, the streamlines have the geometry depicted in that figure. Alternatively, A (y) could be chosen so that $\psi^{(o)}$ vanishes on the western side of the basin; at this stage it is not better than the other choice (on theoretical grounds) and is worse on observational grounds. With $A(y) = X_2(y)$, the predicted trans-

port of fluid is roughly half of that which has traditionally been inferred from observation. It is uncomfortable to note, however, that some recent observations by P. Niiler (not yet published) suggest that the transport is really many times larger than that given by $\psi^{(o)}$.

Aside from the magnitude correspondence, it is quite clear that the streamline pattern cannot be taken seriously until the theory is modified in such a way that the streamline are all closed within the basin. Such a modification must depend on the retention of the friction term and/or the inertia terms in Equation (7-1). The consequences of retaining only the frictional term are investigated by Munk and Carrier[5] (also see Robinson[4]). For that theory, the streamline geometry has the form shown in Figure 7-4. (This is why, in the " $\psi^{(o)}$ theory," $A(y) = X_2(y)$ is chosen.)

But, only if an outlandishly large value for the eddy viscosity is chosen does the western current width correspond to observation.

Even with that choice, the path of the northern part of the current seems not to agree with observation, but this conceivably could be due to an inappropriate simplification of the geometry. Charney[6] studies the flow in the southern part of the basin under the hypothesis that only the inertial term of Equation (7-1) is needed to augment the two rightmost terms. In Reference 4, this idea is carried further and friction is allowed to play its role.

From that theory one finds the following disconcerting results. It is shown that in the absence of friction, any narrow current which travels up the boundary $X = X_1(y)$ cannot proceed beyond the point at which the derivative with regard to y of the right-hand side of Figure 7-3 is zero. That is, no inertial current can go beyond the latitude at which the curl of the wind latitude at which difficulty occurs is the latitude at which currents in the Atlantic and the Pacific seem to depart from the boundary and head in a more easterly direction. Only by a fairly strong faith in coincidence can one avoid the conclusion that the analysis is relevant.

However, no one has succeeded in finding a solution of the equations (either analytically or numerically) which is valid over the entire basin, and the present state of this theory is far from satisfactory. The physical nature of the difficulty is easily spotted and is most conveniently described in terms of vorticity transport. Over the whole basin, the wind is adding clockwise vorticity in the amount Sin y (per unit area per unit time). Only along the boundary is any vorticity lost, and that loss occurs through frictional diffusion (turbulent, of course). However, near $X = X_1(y)$ in $o < y < \frac{\pi}{2}$, the velocity profile has the form shown in Figure 7-5, and frictional mechanisms are effective only to the left of X_o.

Thus, if the transfer of vorticity from particle to particle (and particle to boundary) occurred only in such a boundary-hugging layer throughout the basin, there could be no steady state.

Particles on interior streamlines would acquire more and more vorticity and never lose it to the boundaries. It follows that, in order to have a flow which is steady in the large, the fluid particles must find paths such that the needed vorticity transfer can occur.

One set of trajectories which might do the job would be a set which meanders across the basin, shaking off eddies and redistributing the vorticity by this large scale irregular process. Such a stream is the observed one, but a theoretical formalism is needed to describe it. Currently, this author thinks that the instabilities that imply the actual meandering stream lead to irregular motions which, in effect, are equivalent to an "outlandishly" high turbulent viscosity. The adoption of such a frictional description in the region where the intense current prevails ($\dot{m} \, y > \frac{\pi}{2}$) may well be the realistic "steady state" mechanism which will resolve this question.

Carrier[2] studies alternative approach. There, one supposes that the fluid is (acts as though it were) homogeneous above the thermocline and adopts a state of hydrostatic equilibrium below the thermocline. He then analyzes the Ekman layers near the quasi-spherical surfaces, which constitute the top and bottom of the rotating container, and looks for a flow between these layers which is independent of z, the coordinate parallel to the Ω. When this is done and when a new set of coordinates which is equivalent to those of the foregoing theory are introduced, one obtains an equation for the stream function ψ, which is precisely Equation (7-1) again except that β is now a multiple of the rate of change with latitude of the depth of the basin measured in the z direction.

If the bounding surfaces were spherical with centers at the gravitational origin, β would again be $2 \, \Omega \, \cos \vartheta$; but this, of course, is not the case. The motion of the fluid relative to the rotating system implies a free surface position (pressure distribution), which is not quite one of the spherical surfaces mentioned above. Furthermore, this discrepancy in position implies that, in order for the deep water to be in hydrostatic equilibri-

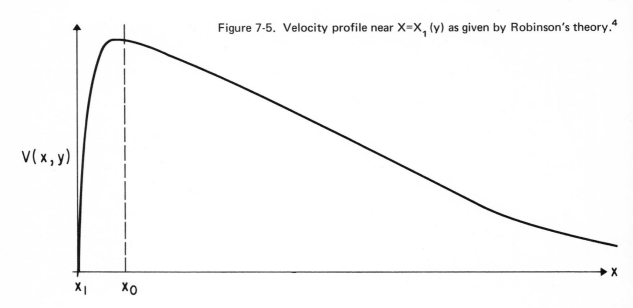

Figure 7-5. Velocity profile near $X=X_1(y)$ as given by Robinson's theory.[4]

um, the thermocline is heavily displaced from its nominal spherical geometry. It turns out that β is much smaller under this approach than it was in the former one, and the transport predicted is larger.

To decide between these two approaches requires a decision as to whether the fluid acts essentially as a homogeneous fluid or as a highly stratified one. The parameter which, according to section 2 of this chapter, is crucial in that decision depends very much on the ratio of horizontal transport coefficient to the vertical one as well as on the vertical transport itself. The uncertainty in these quantities permits estimates implying fully stratified behavior, homogeneous behavior or anything in between. This is convenient in that it allows one to match any transport he wants but it certainly makes it difficult to be assertive on a truly predictive basis!

Aside from predictions of transport, everything discussed under the first class of theories is equally pertinent under this theory and further discussion is not needed.

Unresolved Questions

The following questions must be answered before one can assert that he understands the midlatitude circulation.

1. Is the circulation essentially a two-layer system in which each layer behaves as though it contained a homogeneous fluid, or does stratification so dominate the flow that an integrated transport theory is appropriate?

2. Does there exist a solution of Equation (7-1) subject to appropriate boundary conditions? Does it describe a flow resembling the observed one (almost surely, no)? If not, is it dynamically unstable (almost surely, yes)?

3. Is a non-steady theory essential to an understanding of the dynamics, or can one use a large eddy viscosity to depict the long-time average implications of the meandering stream?

4. What facets of the topography are important for the prediction of the trajectory of the intense current?

5. Does bottom friction take out a substantial amount of vorticity—particularly under the current—or is a horizontal friction model adequate?

These cannot be answered now—the details of past efforts to answer them are given in the papers listed in the References.

References

1. Taylor, G. I. "Experiments with Rotating Fluids, *Proc. Roy Soc.*, A100 (1921), 114; and *Cambridge Phil Soc. Proc.*, 20 (1921), 326.

2. Carrier, G. F. "Phenomena in Rotating Fluids," *Proc. of the 11th International Congress of Appl. Mech.*, Munich, 1964, (Springer-Verlag).

3. ___"Some Effects of Stratification and Geometry in Rotating Fluids," *JFM*, 23 (1965), 145.

4. Robinson, A. R. *The Wind-driven Ocean Circulation*. Blaisdell Publ. Co., (1963).

5. Munk, W. H. and G. F. Carrier. "The Wind-driven Circulation in Ocean Basins of Various Shapes," *Tellus*, 2, No. 3 (1950-51), 158.

6. Charney, J. G. "The Gulf Stream as an Inertial Boundary Layer," *Proc. Nat. Aead Sci. Wash.*, 41 (1955), 731.

part 6: norman j. wilimovsky

Elements of Fishery Resource Management

Engineering Aspects of Fisheries Science

Dr. Norman J. Wilimovsky obtained his B. Sc. (1948) and M.A. (1949) from the University of Michigan, majoring in the biological sciences. After working as an associate ichthyologist on the fishery survey of Brazil, he studied at Stanford University, where he received his Ph. D. (1955) based on arctic hydrobiological investigations.

In 1956 Dr. Wilimovsky joined the research division of the U.S. Fish and Wildlife Service in Alaska, where he served as chief of marine fisheries investigations (Alaska). Late 1960 he accepted a faculty appointment at the University of British Columbia, Vancouver, where—from 1963 through 1966—he was director of the Institute of Fisheries. During this period he conducted studies throughout arctic Alaska and the Aleutian Islands. He was granted leave in 1967 to serve as a senior staff member of the U.S. Council on Marine Resources and Engineering Development.

In 1968 he returned to the University of British Columbia, where he continues as professor in the Faculty of Graduate Studies and the Institute of Animal Resource Ecology.

Dr. Wilimovsky has served as a member of the U.S. Atomic Energy Commission's committee on Project Chariot, editing the volume reporting that work. He participated in the biological studies of Project Long Shot at Amchitka. He has served as president of the Alaska Division of the American Association of the Advancement of Science and is a fellow of the parent organization. He was elected

Fellow and is a past governor of the Arctic Institute of North America, as well as a member of several other societies. Dr. Wilimovsky's researches have included work in systematic ichthyology, fisheries, ecology of ice and underwater instrumentation. Among current interests are the development of science policy and technological forecasting in the resource area.

8

elements
of fishery resource
management

Ocean engineers will frequently be either directly involved in projects involving fisheries, particularly their biology, or in projects indirectly related to them. Consequently, it is paramount that engineers have a fundamental understanding of the processes and factors affecting fisheries. It is the purpose of this chapter to convey some of this background.

Fisheries science, as contrasted with fishery biology or other more limited concepts, is comprised of a number of segments. The most important of these are the resource, harvest, processing, marketing, legal, social and educational factors. These elements may be considered part of an overall system.

The study of the resource comprises investigations of the species, its sub-populations, the environmental parameters affecting them and the nature of the individual populations themselves, that is, their population dynamics.

The importance of knowledge of the environment and the species and its sub-populations, i.e., systematic ichthyology, should be obvious. Such fundamental knowledge is basic to all other work. Inasmuch as the parameters affecting population

dynamics can change from stock to stock, the nature of such an investigation involves a study of distribution and movement of the population through such techniques as tagging, tracking and examination of fundamental life history as well as through sophisticated numerical analysis of the variability of the species.

There are many factors which affect the distribution and abundance of the aquatic biota occupying a given geographical area. The most important factors appear to be the aquatic climate, particularly as related to temperature and salinity; the general nature of the habitat, that is, inshore-shelf inhabitants, benthic forms, pelagic life and distance from fundamental biotic source or "mother fauna."

Most fish groups are thought to have evolved in the inshore habitat, and with time the various species further differentiated and populated more specialized habitats as the pelagic and deep sea. In the course of evolution, the tropic areas evolved the greatest number of species per unit area, the temperate areas fewer and the polar waters the least. In terms of abundance, however, there appear to be fewer individuals of each species living in the

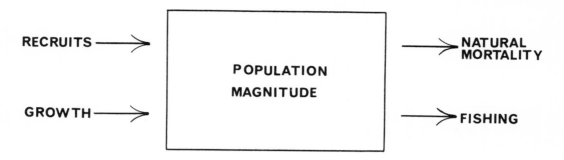

Figure 8-1. Factors controlling fish population biomass.

tropic seas as compared with those living in waters of higher latitudes (or in water masses characteristic of those regions).

Numbers of individual species, that is, their abundance, is determined by habitat "space" within the biogeographic realm and available nutrients. Space as thus defined is a complex feed-back function of the kinds of animals and plants present and aquatic geography, while food available is a function of the aquatic climate and type of life at each trophic level.

The population dynamics of a single species of stock can be considered in the concept of inputs and outputs of a "black box." Biologically, four factors control the weight or biomass of a fish population (Figure 8-1).

Increases in populations are due to births or recruits and growth; the deletions are due to natural deaths or mortality, and fishing. Although much effort has gone into elucidating these processes, the black-box concept is not without meaning or unrealistic, as we have only rudimentary knowledge of the effect of density, food availability, and other dynamic features of the environment on a population and their effect on the total biomass. In its simplest form, the process may be considered as follows: from birth onward, the numbers of fish decrease due to natural mortality. From birth onward, the weight of an individual fish increases due to growth. The potential total weight of a population is the result of these two factors.

Maximum biological yield may be achieved at a time of greatest population biomass, that is, at the point of greatest area under the two interacting rate curves. If yield is considered as some fixed proportion of the population at a given time, it is clear from Figure 8-2 that an early harvest would be the product of many numbers times low individual weight and would produce a low total yield (Point A). Harvesting late (Point C) also results in low total yield due to reduced numbers of individuals though of greater individual weight. At some intermediate point (Point B), one would obtain the maximum yield.

The objective of fishery management in *biological* terms is a sustained harvest at optimum mean yield. These objectives are modified by economic requirements and sometimes by political or social values.

Management at the biological level attempts to influence the inflow and outflow from the population. One attempts to leave optimum numbers for spawning stocks by limiting catch, or through improving natural spawning areas or by creating artificial ones. Attempts to enhance yield through artificial propagation or improvement of the environment have also received attention. Natural deaths may be reduced conceivably through predator controls, but the only parameter through which man has any real control is fishing.

Many techniques for regulating fisheries have been employed and there are innumerable regula-

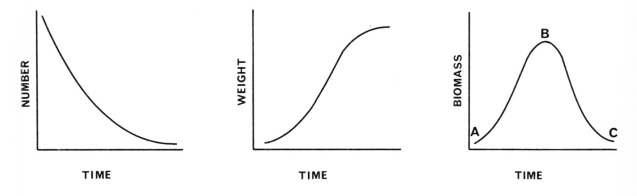

Figure 8-2. Relation of growth in numbers and weight to magnitude of biomass.

tions in force (see example in Table 8-1) but all techniques reduce to means of controlling fishing effort and/or size of fish at first capture.

Table 8-1
Examples of Regulatory Methods

Type of Regulation	
Fishing Effort	Size at First Capture
Type and size of vessel	Mesh or hook size
Number of units of gear	Seasonal closures
Quotas	Fishing area limits (to
Closure of fishing area	avoid nursery grounds)

A number of mathematical models have been evolved for predicting optimum yields. One of the simplest but most practicable considers that there is an upper limit to population size (K) and that the population (N) increases at a rate (r) proportional to the difference between its size at that time and the upper limit. One can consider the population growth as the difference between birth and death rates:

$$\frac{dN}{dt} = rN \frac{K - N}{K} \qquad (8\text{-}1)$$

$$\frac{dN}{dt} = rN - aN^x - cN \qquad (8\text{-}2)$$

It is convenient to rewrite the formula in the second form: where c is the proportion of removals due to fishing. This formulation generates a parabola when numbers are plotted against stock size.

The effect of fishing on stock is obviously to reduce its size. At low population levels, gains from recruitment and growth will exceed losses from natural mortality. If the catch is equal to the surplus, the stock size will not change. Any catch greater than the production due to growth and recruitment will decrease the stock. In a logistic population, it will be apparent that sustainable yield is small at very high stock levels because natural mortality is just less than growth and recruitment. Similarly, yield is small at the low stock levels where the value of gain from growth and recruitment is small. The greatest sustainable yield can be taken at some intermediate stock level with moderate fishing effort. As indicated in Figure 8-3, this occurs at the inflexion point in the rate of

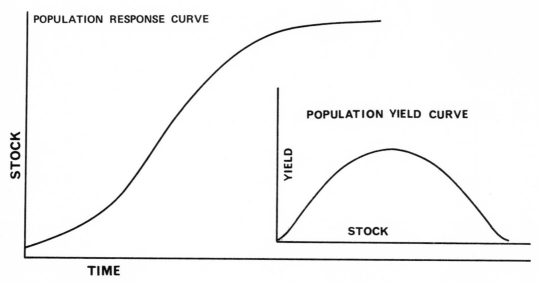

Figure 8-3. Population response (growth) curve and concomitant parabolic relation of yield to stock (after Ricker[5]).

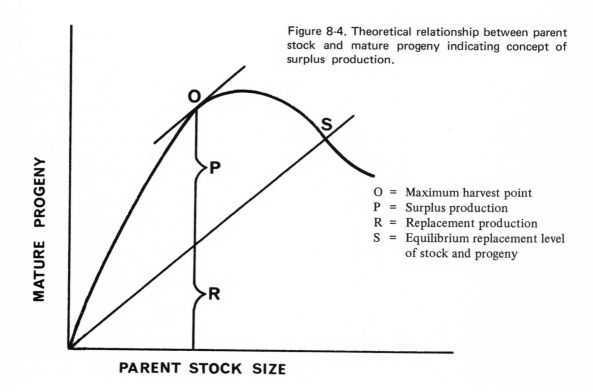

Figure 8-4. Theoretical relationship between parent stock and mature progeny indicating concept of surplus production.

O = Maximum harvest point
P = Surplus production
R = Replacement production
S = Equilibrium replacement level of stock and progeny

population growth and where the population size is equal to $r^2/4a$ for a parabolic function. Unfortunately, most natural populations are asymmetrical in growth structure.

A more analytic approach to prediction of yield is the age-specific model developed by Beverton and Holt.

$$Yw = FRW_\infty e^{-M(tc-tr)} \sum_{n=o}^{3}$$

$$\frac{\Omega_n e^{-nK(tc-to)}}{F+M+nK} (1 - e^{-(F+M+nK)\lambda}) \qquad (8\text{-}3)$$

where Yw = yield in weight
 F = fishing mortality coefficient
 R = number recruited at age tr
 W_∞ = asymptotic weight of individual
 M = natural mortality coefficient
 tc = age at first capture
 tr = age at recruitment
 K = catabolic growth coefficient
 Ω_n = summation variable
 to = adjustment for origin of growth curve
 λ = length of fishable life span

As is apparent, this formulation requires knowledge of a great many biological attributes, some of which may be determined only with much research effort and yet contribute little to the total precision and accuracy of the prediction. This is not a reflection on the nature of the model but the nature of the statistics usually available to the resource manager. There can be no question that better management could result from age-specific models than from the logistic models, but often the "noise" in the system swamps the "signal."

Examination of most models will make clear that the number of recruits is a critical determinant to stock size. Much effort has gone into attempting to define the relationship between parent stock size and resulting offspring. (See Figure 8-4.) One theoretical relationship has been developed at some length by Ricker.[5]

As in the previous logistic curves, the population tends to stabilize at some point S. A maximum harvest of "surplus" production is available at some optimum point O. Unfortunately, the determination of the parameters of this curve is extremely difficult biologically.

If the controlling parameters of fish yield, age of first capture and fishing pressure are plotted for each yield combination, a series of isopleths can be defined from which the eumetric yield curve can be calculated. This curve, which represents the optimum combination of age of first capture and fishing effort, has no maximum but approaches an asymptote at infinity. Such a curve has little value in the economics of real life and one must transform its axes to cost-value terms (Figure 8-5).

Fishing effort may be equated with cost and yield converted to a monetary value. In the simplest cases, such transformation results in a production curve that has a maximum value and consequently suggests an economic "optimum." However, fisheries are generally a common property resource, and an understanding of some of the consequent economic factors is essential to the analysis.

One of the many implications of the common property aspects of fishery resources is that there is no limit on the number of participants. Thus, in any given fishery the development pattern tends to be similar to that of any other. In the natural state, recruitment and mortality are at an equilibrium. Man then enters as a predator. At the outset, his catch per unit of effort is large, profits accrue and new entrants appear. As their numbers increase, the total catch rises and catch per unit of effort declines. Almost invariably the fishery is harvested to the point of profitlessness. The total cost of production is equal to the total return from the fishery. The economic "rent" is completely dissipated among the units of labor, and capital is substantially in excess of that required to physically harvest the stock.

Significantly, because of the common property factor, no single enterprise can justify individual effort for resource conservation. Any effort must be communal by definition, but since fisheries tend to be international in scope and the interests of each nation may not be similar, agreement becomes difficult and multi-national conservation

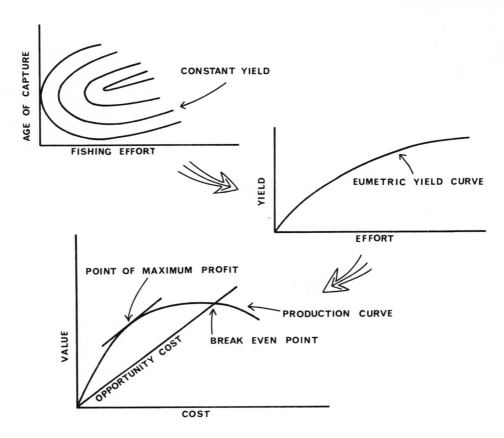

Figure 8-5. Transformation of biological yield curve to production function.

measures have tended to be ineffective in competitive situations. Such conservation measures as are effective tend to be directed toward limiting the efficiency rather than the numbers of participants. Unfortunately, most fishery regulations in force today favor the increase of inefficiency. Some attempts have been made to use regulatory action to reduce effort and consequently the cost function so as to increase the total good to the fishing community. With few exceptions, efforts to date have been noteworthy by their lack of success.

The general economic evolution of a fishery may be shown by a series of phase diagrams (Figure 8-6). It is obvious that without limitation of effort, the aforementioned profitless condition is a natural development in a mature fishery.

Biological and economic factors are not the only constraints on these production functions. For example, the optima dictated by the economic and biological curves are time-dependent, but the size of the fish at some "optimum" time may not be the most desirable to the consumer or possibly to the technology of the processing system. Such constraints may indeed affect the concept of maximum "biological" or optimum "economic" sustained yield.

Finally, it is to be emphasized that the foregoing has been based on the concept of a single-species model, whereas in fact most fisheries harvest several species. The multi-species models developed to date are largely theoretical and pose many practical problems. Even with simple logistic mod-

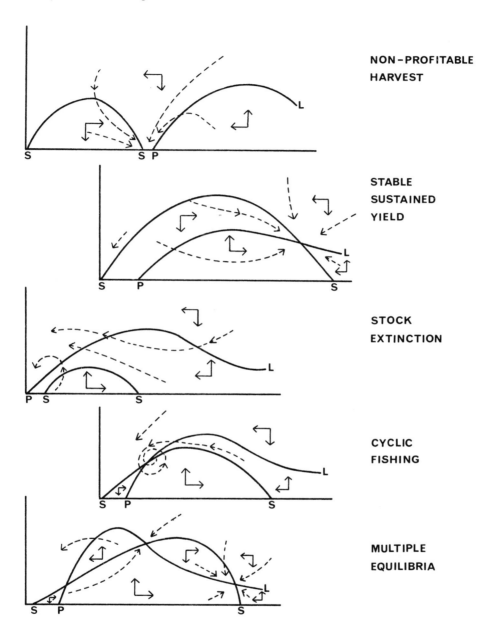

NON–PROFITABLE
HARVEST

STABLE
SUSTAINED
YIELD

STOCK
EXTINCTION

CYCLIC
FISHING

MULTIPLE
EQUILIBRIA

Figure 8-6. Phase diagrams showing relation of stock and investment to harvest (after Quirk and Smith[4]). SS—stock-harvest curves; PL—capital movement curves.

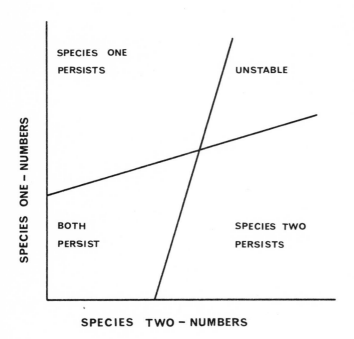

els (Figure 8-7), relatively small changes in numbers at critical stock combinations can bring about drastic shifts in population composition. In multi-species fisheries, as Larkin[2] and Paulik, et al.[3] have pointed out, "overfishing" on one member of the complex may not only be desirable but necessary to obtain optimum yield from the aggregate species complex.

In summary, stocks can be managed for biological, economical or social-political benefits or some mix of these objectives. The interrelationships of the components of the entire fishing system are such that management becomes a series of trade-offs. Cost-benefit analysis is empirical or emotional rather than quantitative. The situation in developed nations is further complicated since recreational time can become more important than extractive use of the sea. As such, sport and recreation add still other considerations to the already difficult problems of multiple-use resource policy (Wenk,[6]).

Key biological questions in long-range fishery management concern stock-recruitment relation-

ships and the nature of the interaction among harvested species. (There is a host of methodology problems not referred to here.) The impossibility of adherence to maximum sustainable yield for a single species when two or more ecological interdependent forms are being exploited should be recognized. The common property nature of a resource has materially contributed to the generally uneconomic condition in fisheries. Management agencies have been thus far largely unable to cope with the fundamental problems of limiting entry. National fishery policies are generally only weakly developed.

The reason for uneconomic fisheries and the solutions to overcome the problems are known. The overriding questions of fisheries science today are how to implement the solutions so that changes to the existing social and economic momentum will be such that the alternative scheme of management will be fair to all involved.

The engineering implications of these factors will be treated in Chapter 9. They largely lie in one factor of fisheries science not treated here, that is, the harvest segment of the total system. Nevertheless, impact of the ocean engineer directly on the biological and economic problems outlined above should not be underestimated. The entire area of instrumentation, the problems associated with monitoring environment and biota, as well as technical questions in processing and handling, await engineering attention.

References

1. Beverton, R. J. H. and S. J. Holt. "On the Dynamics of Exploited Fish Populations," *Fish. Invest., U. K. Ministry Agric. Fish & Food,* Series 2, 19 (1957), 1-533.

2. Larkin, P. A. "Interspecific Competition and Exploitation," *Jour. Fish. Res. Bd. Canada,* 20 (1963), 647-678.

3. Paulik, G. J., A. S. Hourston, and P. A. Larkin. "Exploitation of Multiple Stocks by a Common Fishery," *Jour. Fish. Res. Bd. Canada,* 24 (1967) 2527-2537.

4. Quirk, James P. and Vernon L. Smith. "Dynamic Economic Models of Fishing." *Economics of Fisheries Management—A Symposium.* H. R. MacMillan Lectures in Fisheries UBC, (1970) 3-32.

5. Ricker, William E. "Handbook of Computation of Biological Studies of Fish Populations," Bulletin 119, *Fish. Res. Bd. of Canada,* 1958.

6. Wenk, Edward. "Federal Policy Planning for the Marine Environment," *Public Administrative Review,* 28 (1968), 312-320.

9

engineering aspects of fisheries science

The purpose of this contribution is to explore the role of conventional engineering technology, as well as that of the industrial, military and space fields as to their application and use for improving existing or new concepts of harvesting marine life. Engineering applications fall into three orders of magnitude: improvements to existing equipment, development of new hardware for existing concepts and the engineering of completely new concepts. Examples will be drawn from that sector of fisheries science not previously treated, i.e., the harvesting segment.

Search and Identification

Fishing is essentially a hunting activity. As such, it comprises a strategy of three phases: search, identification and capture. Primitive as well as contemporary techniques of search comprise visual scouting, looking for such things as sea-bird activity, phosphorescent patches—brought about by fish activity at night, listening for sound produced by fish or locating particular environmental areas such as limiting temperatures where it is known that certain species are confined. Detection and location of fish schools can be carried out

aerially, at the surface, or from beneath the surface.

Light-aircraft aerial reconnaissance for fishes has been employed for a number of years. Its success is largely dependent upon the depth of the species, as well as the depth of the water, the transparency of the water and the state of the air-water interface. Clearly, contrast is the significant factor in detectability. Occasionally, the species itself is never seen. Only the interaction between a surface-feeding school and the surface, causing ripples or otherwise a disturbed configuration, is visible. Because of the nature of the water surface, background scatter is such to make radar techniques ineffectual for location of fish populations. Recent tests by the Bureau of Commercial Fisheries in collaboration with NASA using high-flying aircraft provide evidence that dense concentrations or extensive shoals of fishes perhaps can be detected by environmental satellites. However, proceedings of recent remote-sensing conferences provide no "hard" evidence.

Infrared detection has some potentiality for the location of fish stocks. Such radiometric techniques measure surface temperatures and really define the environmental area or disturbed water occupied by the fishes. They do not detect fish per se. The applicability of this technique on a broader scale is limited by the fact that radiometric techniques measure temperatures only at the water-air interface.

Penetration of light through the air-water interface is small except in the blue-green region of the electromagnetic spectrum. Laser offers a means of using extremely narrow band widths in the region of the best "window." Theoretical study implies that even under ideal situations, the use of laser beams would allow reconnaissance of the sea to depths of only 100 meters. To my knowledge, such tests have not been performed.

Obviously, any search, whether by air, surface or submarine, depends upon search patterns. The factors and constraints have been developed at length by operation research specialists in determining effective techniques for anti-submarine warfare. Such procedures are generally most useful in team operations and these are only little developed in fisheries as a whole.

Surface detection and location systems are essentially those techniques described for aerial use except that the sensor is brought down to the sea surface. Even here, every attempt is made to raise one's elevation. For example, the use of tethered balloons increases one's altitude over the deck. The only advantage of surface detection is that the variety of sensors is greater simply because of the nature of contemporary instrument development and the relative ease by which anomalies can be detected at the water-air interface.

The key to fish location and detection lies in the animal's habitat itself, that is, through the use of underwater detection systems. Sub-surface methods of detecting biological resources fall into two general categories: sound methods and non-sound methods (the latter being facetiously referred to as being "un-sound" by non-acoustic workers). Both methods can be employed in either an active or a passive sense. The technological development of sonar has been largely dependent upon military R & D. By its very nature, military sonar development tends to filter out the "noise" which is the very thing the fisherman seeks to

identify and quantify. Consequently, in spite of the enormous developmental cost, it appears that sonar technology for biological purposes must be reengineered. Present systems permit detection of biological populations at ranges of no greater than 15 nautical miles. Pulse coding and doppler techniques in conjuction with side-scanning will allow some refinements, particularly in determining movements of fishes in terms of both direction and speed. Range of passive reception of noise produced by animals is dependent upon the frequency of the sound as well as its magnitude.

There is a real need in the sonic field for high resolution short-range equipment which will allow identification of targets with relative ease once a population is detected. This bears on the all-important problem of target identification. The range at which identification is possible is presently considerably less than the range of detection. Biological and physical data on the patterns of fish shoals in three dimensions would greatly facilitate the remote identification. Just as the field ornithologist can recognize, within a given number of types, the species of bird by the manner in which it flies, so fish schools also exemplify patterns in swimming and schooling which can be recognized. Remote techniques of identifying patterns are within the limits of contemporary technology. An unmanned active sonar interrogator, in either a mobile or fixed submarine buoy, could compare the pattern of a passing fish school with a library of likely species types stored in its computer memory. Identification of sound produced by fishes by the sonic signature is likewise feasible. Both of these developments require considerable engineering perfection before they can become practical realities.

The sonobuoy, developed during World War II for the location of submarines by aircraft has been employed for the location and tracking of fish schools. Essentially, it is a hydrophone attached to a radio transmitter with suitable receiving gear on board the aircraft. By employing several instruments simultaneously, direction and speed of an underwater target can be determined. Fishermen usually lay out a pattern of such buoys and monitor them with existing radio equipment. It should be noted that fishermen render the scuttling device inoperable.

Another exciting potential for the future is the use of acoustic holography. Sonic holograms obtained from sonobuoys could materially revolutionize aerial or surface search and location techniques, as the 3D holograms would provide an estimate of the magnitude of the school as well as offer significant clues for its identification.

Non-sound methods of location and identification of biological resources are largely speculative. One possible visual means is the laser, particularly of the proper wave length. Scuba evidence to date suggests some promise for this technique. Even though it has been used at only very short ranges, the backscatter is much reduced over ordinary light. However, it is likely that shipboard range would be limited.

Perhaps the technique of most probable success would be the detection of resources on the basis of their odors. There is biological evidence for believing that fish can detect extremely small quantities of substances in their environment. Much more investigation is required, but if the basis for olfactory-sensory patterns can be detected, it would be technically possible to develop recording spectrophotometers for tracing organic odors in the sea. With accrual of information on fish odor, it would be possible to identify fish schools by spectrophotometric means. An obvious ramification would be a dispersal of "a distasteful" odor to force or guide fish along a given path or the reverse technique to attract fish to a given point. There is considerable evidence that predators do make use of this principle to "home" in on odor and offal given off by fish schools.

The actual identification of any located stock must be either visual, behavioral or sonic. As previously implied, comparison of a monitored sense pattern with those in a computer memory is technically feasible and offers the ocean engineer an enormous horizon for development.

Once detected and identified, it is important to determine the magnitude of the resource. Interestingly enough, work in this area has progressed much farther than in identification. A number of techniques of interpreting echo returns from sonic devices into equivalent densities of fish stocks have been considered. One which has passed from the theoretical to the applied stage is that used in the United Kingdom (Cushing[1]). The actual analysis is still largely manual but procedures for automation should be comparatively simple.

It will be obvious that search efficiency could be materially increased by combining the flexibility of aerial survey techniques with the range of underwater sensors available. It would be highly desirable to push for a breakthrough in such a procedure, as in areas where aircraft and reconnaissance vessels are currently employed in scouting for a combined fishing fleet, efficiency is very high. Such a procedure is presently in operation off Iceland. Possibly the future would see helicopters using "dipping sonar" for a sensor to obtain sonograms over large areas.

Capture

The second activity of our harvest system or fishing strategy is the capture operation. Many classifications have been proposed and are used for the several types of fishing gear employed in the capture of fish. One means suggests that all contemporary fishing gear can be divided into three categories:

Snares —fish caught directly through their own behavior

Lures —attraction of fish for their subsequent capture

Pursuit—fish caught by force such as surrounding nets and trawls.

The gill net is a good example of a snare; the longline as used for tuna, with its bait or jig are examples of lures. The purse seine and trawl may be categorized as pursuit gear. Combinations of these techniques have been used, and of course other classifications could be applied. There are several good reviews of various types of gear (Davis[2], Kristjonsson[4], Sundstrom[6]).

It is apparent that most of these systems depend upon the nature of the environment and the inherent biological characteristics of the fish to provide concentrations or aggregations sufficiently dense to make capture worth the effort. Thus, contemporary gear is only useful for taking those types of fish that we are currently harvesting. Certainly there is room for expansion of present catches using such gear. Most of our fisheries are concentrated on peripheral sea areas or on the upper-most layers of the pelagic region. The potential harvest of deeper waters or mid-water layers has yet to be realized because of the incapability of the present gear to effectively fish these regions. If man is to fish other habitats such as the mid-open waters or take fishes "lower" in the food web, it will be necessary to devise more efficient straining systems, or develop methods which will cause fish to aggregate and therefore become more easily captured.

Most development of fishing gear and improvement to equipment used today has been an empirical accomplishment. Some of the laboratories have employed studies on visual acuity of fishes in attempts to make nets less visible, i.e., less likely to be detected, as well as towing tank tests to measure and reduce drag of nets. Nevertheless, the analytical approach is missing. Very few data exist on the mathematical modeling of net efficiency or web straining capability. The varied problems offer a wealth of investigations for engineering talent.

There is a vast array of improvements that could be made to existing gear with suitable engineering development. The same statement can apply to ancilliary equipment. For example, improvement could be made in the materials themselves, both in strength and weight. There is a particular need for line gear of electrical conducting material so that the fish could be stunned upon capture, thus allowing for more rapid retrieval.

Even such apparently simple devices as line coilers are lacking on most fishing vessels. The development of a tapered cable of neutral or even positive buoyancy for fishing at intermediate depths would be a material advance to the fishery. There is a great need for communicating to the fisherman the precise position and location of the nets in relation to the fish. Because of the ship's roll,

there is even an ambiguity of the amount of fish available on or near the bottom (Figure 9-1).

Improvement of otter boards or other hydrodynamic net-spreading devices would seem to be an obvious technological advance. The present nature of communication links leaves much to be desired and an accurate acoustic link between the capturing unit and control ship is still to be developed. The combination of lights, pumps and electricity has been explored only to a limited extent (Kristjonsson[4] and Vibert[7]). The effectiveness of such combinations has been demonstrated not only in the Russian kilka fishery but in recent work in the Gulf of Mexico in capturing shrimp in the daylight hours.

The wealth of technological problems involving fishing gear and equipment cannot be overemphasized. Why it has not attracted more engineering interest is beyond comprehension.

Another area directly related to the problems of the fisherman involves cost and weight of vessel construction. Whether fishing boats can be "generalized" and built on the module principle with standardized replacement parts remains to be determined. Much of the problem in the United States is due to the archaic law promulgated in 1792 to protect the U.S. ship building industry which requires a fisherman to use U.S. hulls in order to land fish at U.S. ports. Nevertheless, U.S. engineering technology certainly should be able to make use of the spin-off from other technological areas and employ light, strong materials for vessel and vessel machinery construction. For example, the ferro-cement means of hull manufacture looks most promising.

To advance from a hunting technique to an agricultural stage, the fisherman must be able to confine his "herd." Theoretically, fish can be guided and confined on "marine ranges" using fences or barriers of air bubbles or chemicals, ideally of the filter-bridge type that would let desirable species in and keep the undesirable out. One must be able to move either the fish to the food sources or vice versa. These management concepts are not beyond the reach of biology but require an enormous backup of engineering technology to become

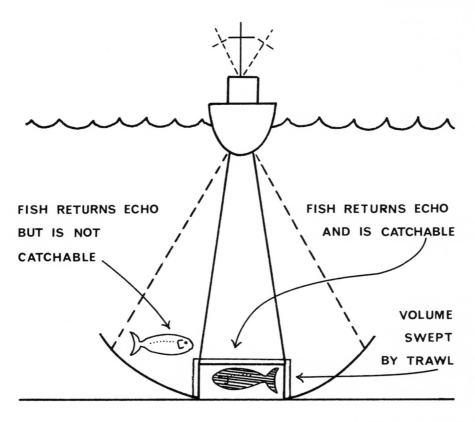

FISH RETURNS ECHO
BUT IS NOT
CATCHABLE

FISH RETURNS ECHO
AND IS CATCHABLE

VOLUME
SWEPT
BY TRAWL

Figure 9-1. Ambiguity of acoustic detection due to effect of ship's roll (after Eddie[3]).

a reality. For example, the development of a cheap, flexible and strong perforated hose whereby air bubbles or water containing a particular chemical could be used to produce a curtain or barrier would find use today in several inshore fisheries.

Another area which has received much attention and, unfortunately, much popular publicity is the artificial propagation of stocks or aquaculture. The development of such "animal husbandry" for fishes should certainly be encouraged in every way. Genetic manipulations of stocks could increase productivity. There are many aquacultural activities extant today and freshwater forms have long been grown in ponds. This subject has been reviewed in a report by Ryther and Bardach[5]. Pi-

lot scale experiments with shrimp, pompano, plaice and other marine species are promising. Nevertheless, in spite of the panacea promised by the popular press, aquaculture as an economic reality lies a considerable distance in the future. It is currently profitable only for high-quality specialty products, and it will be a long time before organisms such as oyster, salmon, plaice and shrimp can be produced at less than $100 a ton. It is doubtful whether aquaculture will ever make a significant contribution to the total protein needs of the world's hungry. Obviously, it could, and does do much in local instances.

If aquaculture is to advance, automation, along with suitable retaining structures are certainly some of the key inputs. Both require engineering talent and particularly at this stage, accurate cost information in order to evaluate cost-benefit relationships. In some of the installations proposed, the cost of the retaining structure was many times the foreseeable profit from the operation. In another existing situation, the cost of food to feed the fish and its distribution exceeds the value of the protein produced.

This overview hopefully has made clear the types of problems facing the resource manager and fisherman. It should also be clear that an intimate understanding of the resource problem by the engineer is a necessity. Similarly, the resource manager-biologist must have a grasp of the technological capabilities. Until this forum and exchange of viewpoints can be accomplished, little progress is likely. Hopefully, with the development of national fisheries policies currently underway, there will be provision for input from the ocean engineering community. In addition to the direction and goals which should evolve from this science policy review, the author would urge consideration of the concept of critical mass. Fishery engineering and ocean engineering efforts are scattered. Very often one group has no knowledge of the other's activities. Individual efforts and funding are at low levels and progress proceeds slowly along most fronts. It would seem realistic and efficient to concentrate a critical mass of talent from all areas in a relatively few—or if necessary, single center—to meet the needs in this broad area. Some steps have been suggested by the U.S. Bureau of Commercial Fisheries in this direction but to no avail. These efforts are likely to continue coming to naught unless the pressure comes from the engineering community itself. Perhaps there is a method by which ocean engineers could communicate their interests to the National Academy of Engineering. Input from that level would necessarily receive consideration by policy makers.

This brief overview has emphasized only a small proportion of the ocean engineering applications to resource management. Time has not permitted consideration of several of the indirect problems such as the effect of underwater explosions on resources, or the relation of marine mining to other resource development. Regardless of the segment of application, one final point should be considered.

It is to be emphasized that reliability is much more important than elegant design in all phases of fishery engineering. By their nature, fishermen are reluctant to accept change without firm conviction. Only dependable systems can be introduced into the field. In the development of equipment, reliability should be placed first and foremost; simplicity second, and finally, operational economy in terms of space, capital cost and operating life.

References

1. Cushing, D.H. "The Counting of Fish with an Echo Sounder," *Rapp. Proces-Verb. Cons. Intern. Explor. Mer.* 155 (1964), 190-195.

2. Davis, F.M. "An Account of the Fishing Gear of England and Wales," *U.K. Ministry Agric. Fish. & Food,* Series 2, 21, No. 8 (1958), 1-165.

3. Eddie, G.C. "Some Recent Developments in Mechanical Engineering in the Deep Sea Fishing Industry," *Proc. Inst. Mech. Engrs.,* Pt. 1, 178, No. 27, (1965), 743-778.

4. Kristjonsson, Hilmar (editor). *Modern Fishing Gear of the World.* Fishing News (Books), Ltd., London, 1959.

5. Ryther, John H. and John E. Bardach. "The Status and Potential of Aquaculture," *The American Institute of Biol. Sci.* 45, 261, 219 (1967).

6. Sundstrom, Gustaf F. "Commercial Fishing Vessel and Gear," *U.S. Fish and Wildlife Service,* Circular 48 (1957), 1-48.

7. Vibert, R. (editor). *Fishing with Electricity.* Fishing News (Books), Ltd., London, 1967.

richard paul
part 7: shaw

Dr. Richard P. Shaw, professor in engineering and applied science at the State University of New York at Buffalo, is presently on leave under an ESSA postdoctoral research associateship with the Joint Tsunami Research Effort at the University of Hawaii, where he is also a visiting colleague in the Department of Oceanography.

He started his teaching career at Brooklyn Polytechnic in 1956, moving on to Pratt Institute (1957-1962) and to New York State (1962). In addition, he has been a consultant to several engineering firms and has participated in several summer institutes, the latest of which was the NATO Research Seminar in Continuum Mechanics and Related Numerical Methods at the University of Calgary during 1968. Dr. Shaw has approximately 25 papers, etc., either published or accepted for publication.

His educational background includes receiving his B.S. in applied mathematics at Brooklyn Polytechnic in 1954, M.S. and Ph.D. at Columbia University (1955, 1960), where he was a Guggenheim Fellow in the Institute of Flight Structures and NSF Fellow.

10

an integral equation approach to acoustic radiation

The class of problems which shall be considered in this discussion involves the motion of small disturbances in a slightly compressible, inviscid fluid medium and the effect that boundaries have on the motion of these disturbances. Classically, this falls under the area of linear acoustics theory and represents a well-known branch of theoretical physics.[1] Applications of this theory can be found in the two most common fluid media available, i.e., the atmosphere and the ocean. Although the mathematical development is the same in either case, the relevant forms of boundary conditions, ranges of applicability of those approximations made in formulating this linearized theory, etc. are somewhat distinct, and an emphasis will be placed on ocean applications for this seminar. Furthermore, emphasis will be placed on a description of the use of this approach to solve problems as well as on the mathematical foundations of the theory.

The variables of interest in these problems are the excess pressure, p, (i.e., excess over an undisturbed state); the density ρ which is assumed to vary only slightly from the undisturbed density ρ_o; and the velocity of a fluid particle, v.

The usual acoustic equations of motion are derived in many texts,[1] and lead for a homogeneous medium to the well-known wave equation, a linear constant-coefficient hyperbolic differential equation, on pressure, density or velocity. It is convenient to introduce a velocity potential ϕ such that

$$v = -\nabla \phi \qquad (10\text{-}1)$$

The linearized equation of momentum will then be satisfied if

$$p = \rho_o \, \partial\phi/\partial t \qquad (10\text{-}2)$$

Then the velocity potential will also satisfy the wave equation

$$\nabla^2 \phi - \frac{1}{c^2} \frac{\partial^2 \phi}{\partial t^2} \equiv \Box^2 \phi = F(\bar{r}, t) \; ; \; \bar{r} \, \epsilon \, D$$

$$(10\text{-}3)$$

143

where F is some prescribed forcing function, c = sound speed in the acoustic fluid and $\bar{r}$ = position vector.

At this point the group of problems to be considered may be divided into two categories: the transient or time dependent case and the time harmonic or "steady state" (in the sense that the time behavior is specified) case.

The transient case requires some form of initial condition, such as the velocity potential field ϕ and its time derivative at some reference time (usually taken as t=o) to be properly posed. In addition, if boundaries S are introduced such that the acoustic medium is restricted to a particular region D (which may be infinite in extent), boundary conditions are required—usually in the form of a linear differential operator of first order, B, i.e.

$$B\,[\,\phi\,(\bar{r}, t)\,] \ = \ H\,(\bar{r}, t) \quad \text{on S} \qquad (10\text{-}4)$$

Typical boundary conditions [2, 3, 4, 5] are

1. Dirichlet: $\quad\phi\qquad$ specified on S

2. Neumann: $\quad\partial\phi/\partial n\qquad$ specified on S

3. Robin: $\quad\phi + K\dfrac{\partial\phi}{\partial n}\qquad$ specified on S

4. Impedance: $Y\dfrac{\partial\phi}{\partial t} + \dfrac{\partial\phi}{\partial n}\quad$ specified on S

5. General: $G\,\phi + \dfrac{\partial\phi}{\partial n} + Y\dfrac{\partial\phi}{\partial t}\quad$ specified on S

$$\qquad(10\text{-}5)$$

These equations are then sufficient to solve any given problem of this form although the solution itself may be extremely difficult to obtain.

The corresponding time harmonic problem assumes a "steady state" solution ϕ with a time dependence $e^{i\omega t}$. In this case, the governing partial differential equation becomes the Helmholtz equation[2]

$$\nabla^2\,\phi + k^2\,\phi \ = \ f(\bar{r}) \ ; \ \bar{r}\,\epsilon\,D \qquad (10\text{-}6)$$

where $k^2 = \omega^2/c^2$ and ϕ now represents only the spatial portion of the solution. Initial conditions are no longer required although some form of radiation condition must be satisfied by time har-

monic solutions involving infinite domains. The boundary conditions in this case correspond to types 1, 2, and 3 of Equation 10-5. The remaining boundary conditions reduce in this case to type 3 with complex values of the boundary parameter K.

Insofar as the geometry of the problems is concerned, a distinction may be made between exterior and interior problems. In the former the region of interest D usually extends to infinity with some interior surface S, on which a boundary condition is specified while the latter usually involves a finite region D bounded by the surface S. The possibility of a mixed problem where two media of different properties are in contact, separated by a boundary S, may also be considered. For the transient and the time harmonic cases, the exterior problem is frequently a scattering problem—i.e., incident waves, ϕ_w, (frequently taken as plane waves) impinge on a "submerged" obstacle and are scattered back into the field. Another exterior problem is that of radiation from a submerged body which is moving.

To simplify the discussion, only the three exterior problems of transient scattering, time harmonic scattering and time harmonic radiation shall be considered, although the approach to be used applies to any acoustics problem.

In terms of underwater acoustics, problems of transient scattering are of interest, for example, in blast loading of structures floating at a "free" surface or submerged in the fluid medium.[6] Although acoustics theory is limited to small changes of density, a relatively simple calculation will show that significant pressure changes can occur within the range of validity of acoustic theory. For example, the speed of sound c in the acoustic medium is $(\Delta p\,/\,\Delta\rho)^{1/2}$ Therefore,

$$\Delta p \ = \ \Delta p\cdot c^2 \qquad (10\text{-}7)$$

For sea water, $\rho\,g\simeq\quad$ 64 lbs/ft^3 and $c\simeq$ 5000 ft/sec. If $\Delta\rho\,/\,\rho$ is taken as 0.01, a one percent change in density, the corresponding pressure will be $\Delta\rho\simeq$ 3500 psi, which is certainly sufficient to cause damage to many types of structures.

While this does not imply that all transient a-coustic waves will cause damage or that all damage can be treated in terms of acoustic theory, it certainly indicates a range of application of acoustic theory to hydrodynamic problems that is not present in the corresponding atmospheric problems where analytical treatment of damage-producing waves frequently requires nonlinear theories. The time-harmonic scattering and radiation have application in underwater detection and identification, since the shape of the obstacle and the relevant boundary condition determine the far field pressure distribution.[7] However, the interesting problem of reconstructing a scattering or radiating obstacle from its far field pressure distribution is beyond the realm of the present discussion.

While it is clear that the proposed mathematical model falls short of reality in many ways—e.g., random medium properties, gross inhomogeneities, over-simplified boundary conditions—sufficient experimental evidence indicates that even this crude model is valuable in practice.

It is the purpose here to discuss an alternate formulation, (i.e., other than the usual differential equation) of these "wave" problems based on the time-retarded potential integral equation of Kirchhoff.[1,2] This formulation applies directly to the transient case but can be readily adapted to time-harmonic problems where the integral equation becomes that of Helmholtz[1,2] (or, in two spatial dimensions, Weber[1,2]). Of course, these problems are interchangeable in the sense that the time-harmonic solution can be generalized to a transient solution by means of a Fourier Series or Fourier Integral[8], while the transient solution for a unit step wave can be extended to the harmonic case (or any time dependence) by means of a Duhamel integral.[8] In practice, however, it still appears preferable to treat these two cases as distinct.

The fact that partial differential equations may have an alternate mathematical formulation as integral equations has long been recognized, e.g., in the use of Green's functions [2,8] which convert the differential equation—assumed to be valid in some region D—into an integral of values of the dependent variable and its derivatives over the boundary S of that region. With an appropriate Green's function for the particular differential equation, the particular geometry and the particular boundary condition involved, the integral equation becomes a simple quadrature, i.e., an integration of known quantities. While this is undoubtedly an ideal manner in which to solve problems, the number of specific Green's functions which are available is quite limited.

However in this process, an interesting reduction of the problem has occurred in that a differential form which involves *all* spatial dimensions has been put into an integral form which can be manipulated to involve only values of the dependent variable (and its derivatives) on the boundary surface S, thereby reducing the spatial dimensions of the problem by one. This can be a significant economy when some approximate or numerical technique is employed to determine the solution of the problem. The basis of the integral equation formulation to be discussed lies in the fact that this reduction of spatial dimensionality can be effected even if the appropriate Green's function for the particular geometry and boundary condition is not known.

Equally widely recognized has been the use of source and sink distributions (in potential flow, for example) to construct solutions to particular problems.[1] While the method discussed here is not quite the same as this "simple source" method, it should be recognized that such an approach is also available and has long been used, for example, in linearized airfoil theory and rigid lighter-than-air vehicles.

Transient Problem

Here the transient problem is considered first. The formulation of this problem into an alternate integral equation form by means of the fundamental solution or Green's function for an infinite domain was apparently first given by Kirchhoff in terms of time-retarded values of the dependent variable. While there are several forms of derivation of this basic equation, they all involve an inherent restriction on the form of the dependent variable chosen. In particular, the dependent variable must vanish at the boundary (as well as outside) of the domain of influence of the scattering surface in four dimensional space-time. This implies that the dependent variable used must represent the scattered wave field, i.e., the total wave field minus the incident wave, or $\phi_s = \phi - \phi_w$ and, furthermore, the dependent variable must be continuous, although it may have discontinuous derivatives which implies that the incident wave potential field be represented by a continuous dependent variable.

Several derivations of the integral equation formulation are available based, for example, on the direct application of Green's theorem in space-time to the wave equation with a point singularity[9]; the extension of the time-harmonic integral solution based on the Helmholtz equation for a form independent of the frequency[2]; or the application of Green's theorem to a four-dimensional space-time volume bounded by the characteristic cone of dependence of a field point, the characteristic boundary of the domain of influence of the secondary (i.e., scattered) front and the hypercylinder representing the scattering surface in space-time.[10]

The forcing function f will be taken equal to zero for the remainder of this discussion although it can readily be put back into the equation. The resulting integral equation on the scattered potential becomes

$$\phi_s(\bar{r}, t) = \frac{1}{4\pi} \iint_{S_o} \left\{ \frac{1}{R} \frac{\partial \phi_s(r, t_o)}{\partial n_o} + \right.$$

$$\left. \frac{\partial R}{\partial n_o} \left[\frac{\phi_s(r_o, t_o)}{R^2} + \frac{1}{cR} \frac{\partial \phi_s(r_o, t_o)}{\partial t_o} \right] \right\} dS_o$$

(10-8)

where $t_o = t - R/c$ is retarded time, n_o is the outward (from the medium) normal, S_o is that surface region containing all points $\bar{r}_o$ capable of affecting the field point $\bar{r}$ at the time t and R is

$|\bar{r} - \bar{r}_o|$. This assumes that the field point $\bar{r}$ at time t is within the region of integration, i.e., exterior to the scattering obstacle.

There is a similar equation for field points not in the volume of integration, i.e., interior to the obstacle

$$0 = \frac{1}{4\pi} \iint_{S_o} \left\{ \frac{1}{R} \frac{\partial \phi_s}{\partial n_o} + \right.$$

$$\left. \frac{\partial R}{\partial n_o} \left[\frac{\phi_s}{R^2} + \frac{1}{cR} \frac{\partial \phi_s}{\partial t_o} \right] \right\} dS_o$$

(10-9)

While either of these equations are sufficient to determine the solution to the problem with appropriate boundary conditions given on ϕ_s rather than ϕ, it will be generally more convenient to work with the total field.

One method of introducing the total field is to consider this as a "saltus" problem.[11] The scattering surface is taken as a surface of discontinuity or "jump" in the dependent variable ϕ and its derivatives. This may be clearly seen by considering Equations 10-8 and 10-9 as the field point r approaches but does not lie on the scattering surface. Obviously, different values of ϕ_s will be found as the field point approaches from the exterior than those found as it approaches from the interior. If there is a jump in value of ϕ_s across this surface, there will be an equivalent jump in ϕ since ϕ_w is continuous across the scattering surface.

The medium is now considered to fill all space with a surface of discontinuities replacing the scatterer rather than have a scattering surface which excludes a finite region from the entire space. However, a jump in values across this surface is not sufficient to completely determine the value on either side. A reference value must be given. In this case the value of ϕ interior to the scattering obstacle will be required to be zero. In other words, the incident field—which is now assumed to pass through the scattering obstacle without change—is wiped out within the obstacle by the secondary (scattered) field, leaving a zero total field within the scattering surface. Since the jump in total field is equal to the jump in the scattered field, a subtraction of equations on the scattered field, i.e., a subtraction of Equations 10-8 and 10-9 will lead directly to

$$\phi_s = \phi - \phi_w = \frac{1}{4\pi} \iint_{S_o} \left\{ \frac{1}{R} \frac{\partial \phi}{\partial n_o} + \frac{\partial R}{\partial n_o} \left[\frac{\phi}{R^2} + \frac{1}{cR} \frac{\partial \phi}{\partial t_o} \right] \right\} d S_o \tag{10-10}$$

where the values of ϕ in the integrand are those taken on by ϕ as the field point approaches the scattering surface from the "exterior"

These values are, of course, equivalent to the values of the discontinuity in ϕ across S_o, since ϕ is required to be identically zero within the scattering obstacle. Therefore, the integral formulation may now be written in terms of ϕ directly; for $r \in$ D

$$\phi(\bar{r}, t) = \phi_w + \frac{1}{4\pi} \iint_{S_o} \left\{ \frac{1}{R} \frac{\partial \phi}{\partial n_o} + \frac{\partial R}{\partial n_o} \left[\frac{\phi}{R^2} + \frac{1}{cR} \frac{\partial \phi}{\partial t_o} \right] \right\} d S_o ; \bar{r} \in D \\ t_o = t - R/c \tag{10-11}$$

This clearly applies to any geometry defined by S_o and will allow for any given boundary condition to be introduced directly into the integral over the boundary. A corresponding equation for $\bar{r} \notin D$ may be written but with a zero left hand side analagous to Equation 10-9, i.e.,

$$0 = \phi_w + \frac{1}{4\pi} \iint_{S_o} \left\{ \frac{1}{R} \partial\phi/\partial n_o + \frac{\partial R}{\partial n_o} \left[\frac{\phi}{R^2} + \frac{1}{cR} \partial\phi/\partial t_o \right] \right\} dS_o \tag{10-12}$$

If the field point is now allowed to approach the scattering surface S_o there will a contribution of $\frac{1}{2}\phi(\bar{r}, t)$ from the singular integrands at $R = 0$ at any surface point where the boundary slope is continuous. (For sharp corners, the value of the coefficient multiplying ϕ will depend on the angle enclosed by the boundary.) This may be interpreted in terms of Green's theorem as excluding the field point on the boundary by a hemisphere (or other modified partial sphere for a sharp corner) rather than by a full sphere as for a field point lying in the volume of integration. The resulting equation valid for the field on the boundary at a point of continuous slope is

$$\tfrac{1}{2}\,\phi\,(\bar{r},\,t) \;=\; \phi_w \;+\; \frac{1}{4\pi}\iint_{S_o{}^*}\left\{\frac{1}{R}\,\frac{\partial\phi}{\partial n_o} \;+\; \frac{\partial R}{\partial n_o}\left[\frac{\phi}{R^2} \;+\; \frac{1}{cR}\,\frac{\partial\phi}{\partial t_o}\right]\right\}dS_o$$

(10-13)

where the principal value of the integral is to be taken, i.e., $S_o{}^*$ is S_o excluding the point $R=0$.

The form now involves the dependent variable only on the scattering surface, i.e., the spatial dimensions of the problem have been reduced by one.

A typical problem might involve the scattering of an incident plane wave with a unit step change in pressure by an arbitrarily shaped obstacle, although it will be evident that any form of incident wave can be used without substantially changing the solution procedure. Typical physical boundary conditions corresponding to mathematical forms mentioned earlier (10-5) are then

1. Free Surface: (zero pressure)

   $\phi = 0$ and $\dfrac{\partial\phi}{\partial t} = 0$

2. Rigid Surface: (zero normal velocity)

   $\dfrac{\partial\phi}{\partial n} = 0$

3. Inertial Resistance: (pressure proportional
                                                     to acceleration)

   $\phi + K\dfrac{\partial\phi}{\partial n} = 0$

   $\left[\text{or }\dfrac{\partial\phi}{\partial t} + K\dfrac{\partial^2\phi}{\partial n\,\partial t} = 0\right]$

4. Viscous Resistance: (pressure proportional
                                                     to velocity)

   $Y\dfrac{\partial\phi}{\partial t} + \dfrac{\partial\phi}{\partial n} = 0$

5. General   First Order:

   $G\phi + \dfrac{\partial\phi}{\partial n} + Y\dfrac{\partial\phi}{\partial t} = 0$

In all cases these boundary conditions enter directly into the integral of Equation 10-13 and are therefore accounted for in an exact manner even if an approximate or numerical scheme is used for the remainder of the solution.

Although the derivation has been carried out using a dependent variable which is continuous, at this point it is convenient to consider dependent variables, such as pressure or velocity which may have finite discontinuities as in the unit step incident pressure wave problem. This may be accomplished by a differentiation of Equation 10-13 with respect to time (to get an equation on p)[10] or of Equation 10-11 with respect to a spatial coordinate which will be in the normal direction as the field point is allowed to approach the surface S (to get an equation on v).[12] Care must be taken in carrying these differentiations through the integration, since the limits of integration depend on time and space coordinates.

Leibnitz's rule must therefore be used leading to additional terms in the equation arising from the variable limits of integration. This result emphasizes the importance in the original derivation of Kirchhoff's equations of using a dependent variable which was continuous and which vanished immediately behind the incident wave front. Clearly, if this condition was not met, the form of equation would be incorrect requiring the additional terms described above. While these additional terms could have been obtained correctly in the original derivation by a careful consideration of integrals over the characteristic surface representing the boundary of the secondary field, it appears much simpler to obtain them in this manner.

In order to consider these equations more closely, the specific case of a rigid boundary scattering obstacle (case 2 of Equation 10-5) is considered. The governing equation, with the appropriate boundary condition $\frac{\partial \phi}{\partial n} = 0$ included is

$$\tfrac{1}{2}\,\phi\,(\bar{r}, t) = \phi_w + \frac{1}{4\pi} \iint\limits_{S_o^*} \left\{ \frac{\phi\,(\bar{r}_o, t_o)}{R^2} + \frac{1}{cR} \frac{\partial \phi\,(\bar{r}_o, t_o)}{\partial t_o} \right\} \frac{\partial R}{\partial n_o}\,dS_o$$

(10-14)

The physical variable of interest in this case would be pressure which could be obtained by a time differentiation of Equation 10-14

$$\tfrac{1}{2}\,p\,(\bar{r}, t) = p_w + A + \frac{1}{4\pi} \iint\limits_{S_o^*} \left\{ \frac{p\,(\bar{r}_o, t_o)}{R^2} + \frac{1}{cR} \frac{\partial p\,(\bar{r}_o, t_o)}{\partial t_o} \right\} \frac{\partial R}{\partial n_o}\,dS_o$$

where A represents those terms arising from Leibnitz's rule in carrying the time differentiation through the time-dependent limits of integration.

The A terms can always be found explicitly for any form of incident wave and any scattering shape, since they only contain the dependent variable evaluated at a field point on the scattering surface immediately behind the incident wave front. The solution there can contain no diffraction effects (unless the point itself acts as a center of diffraction, i.e., as a sharp corner) and must therefore be identical to the solution for the reflection of the same incident wave by a rigid infinite plane—a well-known "doubling" solution. This same procedure applies to other forms of boundary conditions as well and, in general, terms similar to these A terms, can be found explicitly for any of the boundary conditions mentioned. (See References 3, 4, 10 and 12 for details.)

At this point the question of solution of these integro-differential equations shall be put off until after the discussion of the corresponding time-harmonic case.

Time-harmonic Case

The treatment of the time-harmonic case, unlike that for the transient case, will benefit from separate consideration of two and three-dimensional problems. While the governing differential equation is the same in either case, the fundamental solution introduced to transform the differential form to an integral equation is different. The exterior two-dimensional case uses $\pi\,i\,H_o^{(1)}\,(k\,R)$, where $H_o^{(1)}$ is the Hankel function of the first kind and zero order chosen to satisfy the radiation condition. This gives Weber's solution to the two-dimensional Helmholtz equation.[2]

$$\phi_s\,(\bar{\rho}) \;=\; \frac{i}{4} \int_C \left[\phi_s\,(\rho_o)\,\frac{\partial}{\partial n_o}\,H_o^{(1)}\,(kR) \;-\; H_o^{(1)}\,(kR)\,\frac{\partial \phi_s\,(\bar{\rho}_o)}{\partial n_o} \right]\, ds_o$$

(10-16)

where $\bar{\rho}$ represents the two-dimensional coordinates of the field point, n_o is the outward normal (from the fluid) to the boundary curve C with an arc length parameter s.

Although most of the researchers in this field working on time-harmonic problems appear to prefer to use the scattered potential as the dependent variable, equations in this section shall be put in terms of ϕ, the total potential. It should be clear that the change to the total potential as the dependent variable follows in exactly the same manner as in the time-dependent case. Therefore, Equation 10-16 may also be written

$$\phi\,(\rho) \;=\; \phi_w\,(\rho) + \frac{i}{4} \int_C \left[\phi\,\frac{\partial\,H_o^{(1)}\,(kR)}{\partial n_o} \;-\; H_o^{(1)}\,(kR)\,\frac{\partial \phi}{\partial n_o} \right]\, ds_o \;\;;\; \rho \in D$$

(10-17)

and correspondingly

$$0 \;=\; \phi_w\,(\bar{\rho}) + \frac{i}{4} \int_C \left[\phi\,\frac{\partial H_o^{(1)}\,(kR)}{\partial n_o} \;-\; H_o^{(1)}\,(kR)\,\frac{\partial \phi}{\partial n_o} \right]\, ds_o \;\;;\; \rho \notin D$$

(10-18)

On allowing the field point to approach the scattering "curve," singular integrand contributes $\frac{1}{2}\,\phi\,(\bar{\rho})$ for points lying on for a smooth curve. Again the resulting integral is interpreted as a principal value.

The corresponding three-dimensional problem uses the fundamental solution $\frac{e^{+ikR}}{R}$, again chosen to represent outgoing waves at infinity (for the scattered potential) and therefore satisfy the radiation condition. The corresponding integral equation is[2]

$$\phi_s\,(\bar{r}) = \frac{1}{4\pi} \iint_S \left\{ +\phi_s\,(\bar{r}_o)\,\frac{\partial}{\partial n_o}\left(\frac{e^{ikR}}{R}\right) \;-\; \frac{e^{ikR}}{R}\,\frac{\partial \phi_s\,(\bar{r}_o)}{\partial n_o} \right\}\, dS_o \;;\; \bar{r} \in D$$

(10-19)

Again this may be written in terms of ϕ; for $\bar{r} \, \epsilon \, D$

$$\phi(\bar{r}) = \phi_w(\bar{r}) + \frac{1}{4\pi} \iint_{S_o} \left\{ +\phi_s \frac{\partial}{\partial n_o} \left(\frac{e^{ikR}}{R} \right) - \frac{e^{ikR}}{R} \frac{\partial \phi}{\partial n_o} \right\} dS_o$$

(10-20)

and

$$\bar{r} \, \epsilon \, D, \quad 0 = \quad \text{''} \quad + \quad \text{''} \tag{10-21}$$

As the field point approaches the surface, the singularity again contributes $\frac{1}{2}\phi(r)$; and the remaining integral is again taken as a principal value.

The resulting surface integral equations are then

$$\frac{1}{2} \phi(\rho) = \phi_w(\rho) + \frac{i}{4} \int_C \left\{ \phi \frac{\partial}{\partial n_o} H_o^{(1)}(kR) - H_o^{(1)}(kR) \frac{\partial \phi}{\partial n_o} \right\} dS_o$$

(10-22)

and

$$\frac{1}{2}\phi(\bar{r}) = \phi_w(\bar{r}) + \frac{1}{4\pi} \iint_{S_o} \left\{ \phi \frac{\partial}{\partial n_o} \left(\frac{e^{ikR}}{R} \right) - \frac{e^{ikR}}{R} \frac{\partial \phi}{\partial n_o} \right\} dS_o$$

(10-23)

in the two and three-dimensional cases, respectively.

Since the limits of integration are not time or position dependent but cover the entire body, these equations are of the Fredholm type. No advantage is gained by a formulation in terms of pressure, which is simply $-i \omega \rho_o \phi$. If velocity is sought, the problem could be formulated directly in terms of velocity, since the velocity must also satisfy the wave equation. The only change then would be in replacing the terms ϕ_w by V_w terms representing the incident velocity field.

At this point the overall character of the problems should be discussed. A time-harmonic rigid scattering problem, for example, will involve a non-zero ϕ_w but a zero value for $\partial \phi / \partial n$ in the integral. The radiation problem, on the other hand, will not have an incident wave i. e., $\phi_w = 0$ but will have an inhomogeneous (forcing) function

in the integral representing the vibrating boundary (e.g., a prescribed time-harmonic velocity). There may be some difficulty in solving an exterior scattering problem if the frequency matches a natural frequency for the interior resonance problem with the same boundary condition and geometry, but this can be overcome.[13]

Now that the equations have been formulated, the question shifts to one of determining their solution in any particular problem. Although some analytical solutions can be obtained by this integral equation approach, its primary use has been in the development of approximate and/or numerical solutions.

The approximate and/or numerical techniques described below are those which have been used in some form on the solution of actual wave scattering problems expressed in this integral equation formulation. For the most part, these do not consider the significant questions of stability and convergence, except to mention that they exist. Although some studies have been made on these points,[14] there is clearly room for further effort—particularly for the transient case.

These methods fall essentially into three categories:

1. Direct approximation of the integral could be accomplished by a finite summation (and derivatives by finite differences) by assuming the dependent variable to remain constant over a specified step in a spatial coordinate (and in time for the transient case) with a value equal to that at some specified point in the region bounded by these increments.

    The resulting set of linear algebraic equations is either successive (or very weakly coupled) in the transient case,[10] thereby allowing for a relatively easy solution at each successive time step, since this is a Volterra-type equation, or completely coupled in the time-harmonic case, which is a Fredholm-type equation.[15] This approach appears to be the most widely used and will be emphasized in the remainder of this discussion.

2. An iteration procedure could be employed, using the geometrical acoustics solution as an initial guess. Because of the complexity of the integrations which would have to be carried out, however, this approach might require a numerical integration and differentiation within each iteration, thereby increasing the computational effort beyond that of the direct approach mentioned above. From a theoretical point of view, however, it may be possible to use this approach to investigate questions of stability and convergence.

3. A third approach uses an expansion in a set of complete functions with arbitrary coefficients, which are then determined to give a "best" fit in a least squares or Chebyshev sense. Such functions could be chosen to simplify the integrations required. (Reference 16 does an elastic wave time-harmonic scattering problem by this approach.)

Before specific discussion of any of these methods, a few general statements may be made. There is clearly an alternate formulation to the singular integral equations given above. If the field point is kept out of the fluid region (i.e., inside the obstacle in the exterior problems), the time-harmonic integral expressions Equations 10-18 and 10-21 and the transient case (Equation 10-12) e.g., will have zero left-hand sides. This represents a non-singular integral equation on surface values alone so long as the field point remains away from the boundary and is often referred to as the "interior" formulation as opposed to the "surface" form.[17]

Although some numerical difficulties have been found with this non-singular integral equation approach, it has been used with success—at least in time-harmonic problems. Apparently, even better is an over-determined scheme which employs both forms of the integral equation. This method,[13] CHIEF (Combined Helmholtz Integral Equation Formulation), has been used on time-harmonic problems which presented difficulty by other approaches (e.g., near an interior resonance frequency for exterior scattering problems).

Clearly, any of the solution techniques mentioned above will also apply to these alternate formulations.

The primary difficulty with the non-singular or interior form is that the integral equation becomes one of the first kind, i.e., where the dependent variable appears only inside of the integral. This may also occur in the surface formulation for the free boundary condition where ϕ is zero on the boundary and $\dfrac{\partial\phi}{\partial n}$ is the variable of interest. In this case, for example, the transient equation is simply

$$-4\pi\,\phi_w = \iint\limits_{S_o} \left\{ \frac{1}{R}\ \frac{\partial\phi(\bar{r}_o, t_o)}{\partial n_o} \right\} dS_o$$

$$(10\text{-}24)$$

Although these equations appear somewhat simpler in form than those of the second kind (e.g., Equation 10-12), equations of the first kind are generally less amenable to approximate or numerical solution than those of the second kind.[14] While these difficulties may be resolved, it is frequently easier to transform the integral equation into one of the second kind before a solution is attempted.

Finally, the solution in the remainder of the field needs to be discussed, since this integral equation approach only involves the surface values of the unknowns in the solution procedure. This is, in fact, one of its main advantages in that the dimensions of the problem have been reduced. However, an examination of the original equations—Equations 10-11, 10-17 and 10-20—shows that values in the field may be calculated by simple quadrature over the known surface values, which would have already been found separately by this method.

Several specific problems have been solved by this approach, and the results will be discussed along with an indication of some of the actual numerical procedures involved. The process of non-dimensionalization used in these solutions is not described so as to keep the emphasis on the physical variables of interest, although such a step is really fundamental to the numerical solution of almost any problem.

The transient exterior scattering problem has been treated for the two-dimensional case of an infinite cylinder of square cross section struck laterally by a step pressure wave.[10] This geometry was chosen to examine the diffraction effects, which are centered at the sharp corners. Rigid,[10] free,[12] inertial[3] and impedance[4] boundary conditions were considered in separate papers; and the procedure for the rigid boundary condition shall be described briefly. The boundary condition is $\frac{\partial \phi}{\partial n} = -V_n = 0$ on S and the variable of physical interest is the pressure p which can be found from Equation 10-15, by a direct approximation of the integral by a summation. Choosing the time and space mesh as shown in Figure 10-1 and taking the symmetric case for simplicity, Equation 10-15 is approximated by Equation 10-25 in the box at the bottom of this page.

where

I defines the side on which the field point lies

J defines the side on which the integration (source) point lies

M defines the coordinate of the field point

N defines the coordinate of the source point

T defines the current value of time, i.e., $t = T\Delta t$

L defines the retarded value of time, i.e., $t_o = t - R/C = L^*\Delta t$ where L is the nearest integer value to L^*

$$\frac{1}{2}p\,(I,M,T) = p_w\,(I,M,T) + A\,(I,M,T) + \frac{1}{4\pi} \sum_{J=1}^{4} \sum_{N=1}^{N\,MAX} \sum_{L=1}^{T} \left\{ \alpha\,(I,M,T;J,N,L)\,p\,(J,N,T-L) \right.$$

$$\left. + \beta\,(I,M,T;J,N,L) \left[\frac{p\,(I,M,T;J,N,T-L) - p\,(I,M,T;J,N,T-L-1)}{\Delta t} \right] \right\}$$

Figure 10-1. Geometry for numerical example: infinite cylinder with a square cross section struck symmetrically by a plane acoustic step pressure wave.

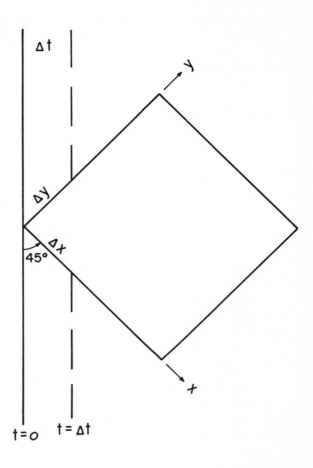

and α,B are coefficients which can be made to depend only on the geometry of the problem (i.e., will not depend on the form of the incident wave). This equation may be applied at successive field points at the same value of time and gives rise to a set of linear algebraic equations which are uncoupled due to the time retardation, i.e., those values of pressure required in the summation are either the particular value sought at the given time and field point (which can then be included in the left-hand side of the equation) or values at earlier times which will have already been calculated.

Corresponding calculations have been carried out for the free boundary condition, the inertial boundary condition (which contains both the rigid and free problems as special cases) and the impedance boundary condition. To simplify the discussion, only two results are shown in Figures 10-2 and 10-3.[3] These indicate the results for an inertial boundary condition, where the boundary is assumed to be made up of unconnected masses each free to move without the elastic restraint of its neighbors. The two limiting conditions of infinite mass (rigid boundaries) and zero mass (free boundaries) are included as special cases.

One result of interest which may be seen from these curves is the relatively wide range of surface mass values over which the free surface boundary

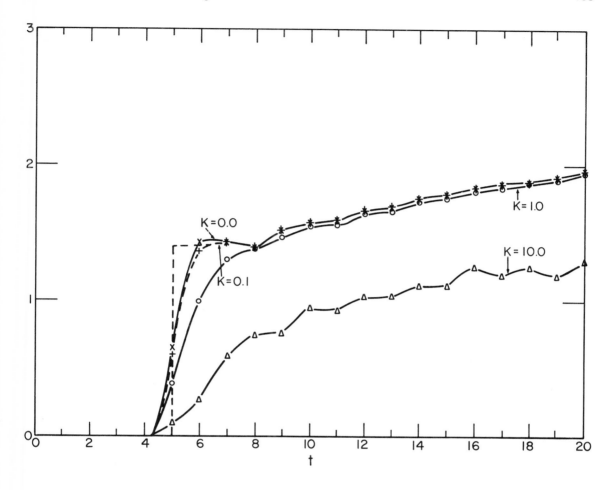

Figure 10-2. Velocity response for semi-free case for square "box," at point m=5 on side 1, θ=45°, T=20, N=10.

condition represents a good approximation to the inertial boundary condition. Taking the value K, which represents the non-dimensional ratio of the mass density of the surface area to the mass density of the fluid per unit volume times the length of a spatial increment (which is of the order of 1/6 to 1/10 of the length of the side of the box), it can be seen that curves for values of K from 0.0 to 1.0 fall very near each other.

The value K=0 represents the free surface boundary condition which can then be used to approximate most submerged structures in a water environment. For example, for K=1.0 a submerged steel shell (of square cross section) would require a thickness to side length ratio of approximately 1/100. This same conclusion has been arrived at in a related problem for another geometry and will be discussed further in Chapter 11. Unfortunately,

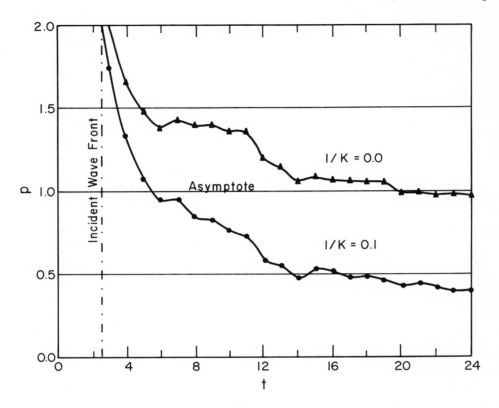

Figure 10-3. Pressure response for semi-rigid case for square "box," at point m=3 on side 1, θ=45°, N=6, T=24.

the corresponding conclusions do not appear to hold for very heavy scatterers (or correspondingly light acoustic media such as air) where the inclusion of finite surface mass effects has a significant effect on the results. $\frac{1}{K}$=0.0 is quite different than $\frac{1}{K}$=0.1, i.e., the rigid boundary condition is not a good approximation in general to the inertial boundary condition for heavy scatterers.

In addition to these, other transient exterior scattering problems solved by this approach include a rigid sphere struck by a plane pressure pulse with a Gaussian distribution[18] and an infinite orthogonal trihedron (three-dimensional corner) struck by a plane step wave in pressure.[19] The former shows good agreement with a corresponding eigenvalue solution while the latter ap-

pears to give a correct solution when examined in terms of time reduced coordinates—i.e., $(\frac{x}{t}, \frac{y}{t}, \frac{z}{t})$— which must govern the form of the solution, since there is no representative length in this problem. The contribution of these papers, in addition to the solution of the particular problems considered, is primarily on questions of subdividing the integral and appropriate step sizes.

The corresponding time-harmonic exterior scattering problem may be treated in a very similar manner except that the resulting set of linear algebraic equations is densely coupled. Since the potential (and therefore the pressure and velocity) is complex, the Helmholtz or Weber equation must really be written in terms of its real and imaginary parts leading to a coupled pair of integral equa-

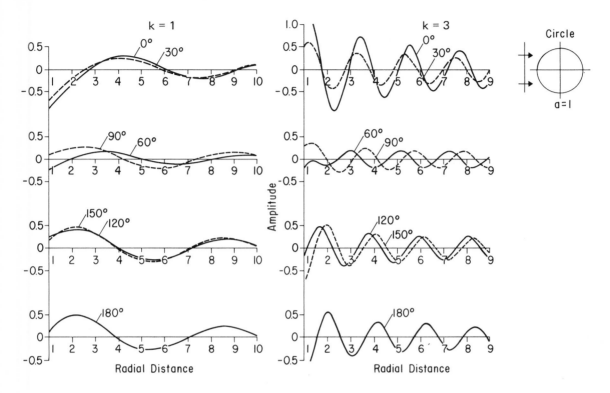

Figure 10-4. Acoustic scattering from a rigid cylinder (from University of California, Lawrence Radiation Laboratory, Livermore, and U.S. Atomic Energy Commission).

tions. Thus, if n field points are used and n pairs of integral equations are approximated, the resulting set of simultaneous algebraic equations is of order 2n. However, for two-dimensional problems, the original integral equations are only one dimensional, thereby reducing much of the arithmetic involved in determining the coefficients of the set of equations in this case.

While this direct solution procedure will not be repeated here, it is very similar to that discussed in the transient case except, of course, for the absence of the time variable and the corresponding reduction in the number of independent variables. Calculations have been carried out for the two-dimensional problems of plane incident step pressure waves scattered by a rigid circular cylinder, a rigid

prolate elliptical cylinder, a rigid oblate elliptical cylinder, a rigid three-leafed rose cylinder, a free circular cylinder, a free prolate elliptical cylinder, a free oblate elliptical cylinder and a free three-leafed rose cylinder.[15] The results for scattering by a rigid and a free circular cylinder were compared to corresponding eigenfunction expansions in the near field (on the surface) and far field and showed good agreement. Results are given in Figures 10-4, 10-5 and 10-6 for some typical examples.

Time-harmonic radiation problems have been solved by the usual direct surface singular integral equation approach and by the non-singular $\bar{r} \notin D$ approach. The former[20] considers rigid vibrations of a sphere and quadrupole-like vibration of a

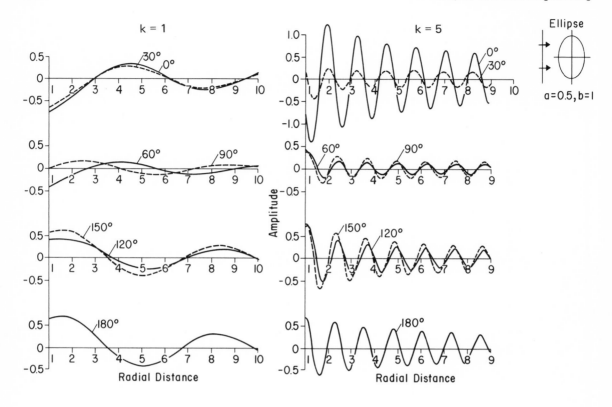

Figure 10-5. Acoustic scattering from a rigid cylinder. (From University of California, Lawrence Radiation Laboratory, Livermore, and U.S. Atomic Energy Commission).

spheroid as examples, although the theory is developed for a general (i.e., not necessarily axisymmetric) class of vibration velocity distributions for surfaces of revolution. By use of a Fourier expansion of the prescribed velocity distribution in terms of the cylindrical coordinate, the resulting integral equations are only one dimensional, again reducing the computational effort, even though they are based on the three-dimensional Helmholtz integral equation (i.e., Equation 10-23). In a manner analogous to that for the scattering case, evaluation of the integral equation at surface points representative of the subdivision used to approximate the integrals leads to a deterministic set of coupled linear algebraic equations.

The latter[17] considers axisymmetric radiators (i.e., surfaces of revolution loaded symmetrically) in which case the integral equation is again one dimensional. The field point, however, is taken to lie at various locations on the axis of symmetry within the surface so as to give a finite number of integral equations based on Equation 10-21 which are then approximated by summations as discussed above. If the number of intervals chosen in approximating the integral is the same as the number of equations (i.e., number of field points used), the system is deterministic. Good agreement for a vibrating spheroid was obtained by this approach with results obtained from an eigenfunction expansion on the surface and in the far field. However, the integral equation is still one of the first kind and is somewhat more difficult to solve numerically than one of the second kind.

An "improved" formulation[13] which combines both approaches, has been used on a uniformly vibrating sphere, a finite right circular cylinder and vibrating rectangular parallelopipeds, with good re-sults in comparison to experiments and other related solutions. This formulation uses the surface integral equation form at those field points defining areas of approximation as in the previous cases and also introduces integral equations for field points lying interior to the scattering obstacle. The net result upon approximation of the integrals by fi-

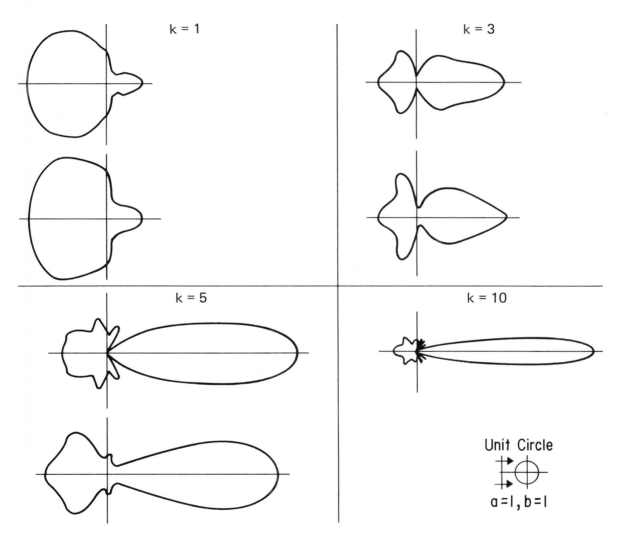

Figure 10-6. Comparison of far field intensities. (From University of California, Lawrence Radiation Laboratory, Livermore, and U.S. Atomic Energy Commission).

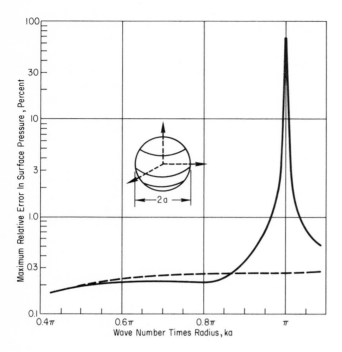

Figure 10-7. Maximum relative error in surface pressure as a function of ka for a uniformly vibrating sphere. Solid line—surface Helmholtz integral formulation; dashed line—combined Helmoltz integral equation formulation with one interior point at center (after Schenck[13]).

nite summations is an over-determined set of coupled linear algebraic equations. This set is then solved by a least squares procedure.

In the examples carried out, it was found that only one additional equation (e.g., for a field point at the center of the radiating object) gave a significant improvement for frequencies close to (or even at) the frequency of resonance of the corresponding interior problem. This region of frequency gave substantial difficulty by the surface and the interior approaches separately. An indication of the improvement is shown in Figure 10-7 for a radially vibrating sphere.

A useful summary of integral equation approaches to time-harmonic radiation and scattering problems has been given[21] and includes mention of other integral equation approaches, such as those based on simple source methods which are somewhat different than those discussed here.

Although only a brief description of this integral equation approach to acoustic problems has been given, the usefulness of this method in the solution of specific problems should be apparent.

To discuss the advantages and disadvantages of this approach, however, other methods of solution to be used as a basis of comparison must also be discussed.

The standard procedure for those geometries to which it is applicable is separation of variables or eigenfunction expansions.[9] This approach leads to an infinite series expansion, frequently in terms of relatively complicated functions, and it is not yet clear whether this approach possesses any real computational advantage over other methods in some of those cases to which it can be applied. Certainly, it is not useful for arbitrary geometries. It might further be added that eigenfunction solutions to transient problems have convergence difficulties at small values of time which is where the integral formulation would be expected to give its best results.

The next standard approach might be the direct approximation of the governing differential equation by finite differences with use made perhaps of characteristics in the time-dependent case.[14] This approach has two distinct disadvantages: (1) the boundary conditions are unlikely to be prescribed on a mesh point and must therefore be further approximated; (2) the entire solution field must be determined simultaneously. The integral equation approach has the corresponding advantages of using the boundary conditions on the boundary and separating the solution on the surface from that in the remainder of the field.

Another widely used method is that of geometrical acoustics, including geometrical theory of diffraction.[22] This method applies to homogeneous and inhomogeneous media and represents a powerful solution technique. Although the method is very simple for very high frequency solutions (i.e., where ordinary geometrical optics theory applies[23]), ocean related problems appear to fall in frequency ranges where higher order diffraction effects are significant. This would then appear, however, to require a more advanced mathematical treatment (e.g., theory of asymptotic expansions) than the integral approach, which seems therefore simpler to formulate, for homogeneous media at least. The integral equation approach does not appear to have been extended to inhomogeneous acoustic media (although some work has been done on related problems in elastic media), and there the "ray" theory of geometrical acoustics appears to be superior.

In summary, the integral equation approach based on either the Kirchhoff time-retarded integral equation or the Helmholtz (or Weber) integral equation appears to present a useful approach to the solution of acoustic wave problems. Its main advantage is a reduction in the dimensionality of the problem to a form involving only surface values. Those values in the remainder of the field may be found from these surface values by direct quadrature. Although some useful analytical results may be obtained by this approach, its primary application has been in the approximate and/or numerical solution of specific problems. Although the usefulness of the approach has been established, there still remain several significant areas of investigation open. These include questions of stability and convergence, particularly in the transient case; problems of minimizing the storage

and/or time requirements (which are relatively large) of these calculations; and extensions of this approach to elastic wave problems (already considered to some extent), acoustic wave-elastic shell interactions, inhomogeneous media, random media, et cetera.

List of Symbols

A	Terms arising from differentiation of integral with variable limits (Equation 10-15)
B	Boundary condition operator (Equation 10-4)
c	Sound speed in acoustic fluid (Equation 10-3)
D	Domain of acoustic fluid
f	Forcing function for harmonic wave equation (Equation 10-6)
F	Forcing function for transient wave equation (Equation 10-3)
G	Boundary condition parameter (Equation 10-5)
H	Boundary forcing function (Equation 10-4)
$H_0^{(1)}$	Hankel function of the first kind of order zero
I	Index in Equation 10-25
i	$\sqrt{-1}$
J	Index in Equation 10-25
k	Wave number
K	Boundary condition parameter (Equation 10-5)
L	Index in Equation 10-25
M	Index in Equation 10-25
N	Index in Equation 10-25
n	Outward normal (from fluid)
p	Excess pressure (Equation 10-2)
$\bar{r}$	Position vector; field point (x,y,z)
R	Distance between field point and source point (Equation 10-8)
S	Boundary of fluid
S_0	Integration surface
s	Arc length in two-dimensional case
t	Time
t_0	Retarded time = $t - R/c$

Δt Time step
T Index in Equation 10-25
$\overline{v}$ Particle velocity (vector)
Y Boundary condition parameter (Equation 5)
α Geometrical coefficient (Equation 10-25)
β Geometrical coefficient (Equation 10-25)
ϕ Total velocity potential
ϕ_w Incident velocity potential
ϕ_s Scattered velocity potential
ρ Density
$\overline{\rho}$ Two-dimensional position vector
ω Wave frequency
∇^2 Laplacian Operator
$\square^2$ Wave Operator$\equiv \nabla^2 - \dfrac{1}{c^2}\dfrac{\partial^2}{\partial t^2}$

References

1. Lamb, H. *Hydrodynamics.* Dover, NYC, 1945.

2. Sneddon, I. N. *Elements of Partial Differential Equations.* McGraw Hill, NYC, 1957.

3. Shaw, R. P. "Diffraction of Acoustic Pulses by Obstacles of Arbitrary Shape with a Robin Boundary Condition," *Jour. Acous. Soc. Am.,* 41, No. 4 (April 1967), 855-859.

4. _____. "Diffraction of Pulses by Obstacles of Arbitrary Shape with an Impedance Boundary Condition," *Jour. Acous. Soc. Am.,* 44, No. 4 (October 1968), 1062-1068.

5. _____. "Scattering of Plane Acoustic Pulses by an Infinite Plane with a General First Order Boundary Condition," *Jour. App. Mech.,* 34, No. 3 (September 1967), 770-772.

6. Testa, R. and H. H. Bleich. "Dynamic Response of a Floating Rigid-Plastic Box," ASCE, *Jour. Eng. Mech. Div.,* 92, No. EM2 (April 1966), 131-151.

7. Kendig, P. M. "Advanced Transducer Developments," *Underwater Acoustics, Vol. II.* V. M. Albers (ed.), Plenum Press, N. Y., 1967.

8. Duff, G. F. and D. Nayler. *Differential Equations for Applied Mathematics.* John Wiley & Sons, NYC, 1966.

9. Morse, P. M. and H. Feshback. *Methods of Theoretical Physics Vol. I and II.* McGraw Hill, NYC, 1953.

10. Friedman, M. B. and R. P. Shaw. "Diffraction of a Plane Shock Wave by an Arbitrary Rigid Cylindrical Obstacle," *Jour. App. Mech.,* 29, No. 1 (March 1962), 40-46.

11. Baker, B. B. and E. T. Copson. *The Mathematical Theory of Huygens' Principle.* Second edition, Oxford Press, 1953.

12. Shaw, R. P. and M. B. Friedman. "Diffraction of a Plane Shock Wave by a Free Cylindrical Obstacle at a Free Surface," *4th U. S. National Congress of Applied Mechanics* (June 1962), 371-379.

13. Schenck, H. A. "Improved Integral Formulation for Acoustic Radiation Problems," *Jour. Acous. Soc. Am.,* 44, No. 1, (July 1968), 41-58.

14. Fox, L. *Numerical Solution of Ordinary and Partial Differential Equations.* Addison Wesley Publishing Co., NYC, 1962.

15. Banaugh, R. P. "Scattering of Acoustic and Elastic Waves by Surfaces of Arbitrary Shape," *UCRL - 6779.* University of California, February 1962; and R. P. Banaugh and W. Goldsmith. "Diffraction of Steady Acoustic Waves by Surfaces of Arbitrary Shape," *Jour. Acous. Soc. Am.,* 35, No. 10 (October 1963), 1590-1601.

16. Sharma, D. L. "Scattering of Steady Elastic Waves by Surfaces of Arbitrary Shape," *Bul. Seis. Soc. Am.,* 57, No. 4 (August 1967), 795-812.

17. Copley, L. G. "Integral Equation Method for Radiation from Vibrating Bodies," *Jour. Acous. Soc. Am.,* 41, No. 4 (1967), 807-816.

18. Mitzner, K. M. "Numerical Solution for Transient Scattering from a Hard Surface of Arbitrary Shape—Retarded Potential Technique," *Jour. Acous. Soc. Am.,* 42, No. 2 (1967), 391-397.

19. Shaw, R. P. and D. F. Courtine. "Diffraction of Plane Acoustic Pulse by a Free Orthogonal Trihedron," *D.I.S.R. Rep. No. 32,* State University of NY at Buffalo, May 1968. Also *Jour. Acous. Soc. Am.,* 46, No. 5 (November 1969), 1382-1384.

20. Chertock, G. "Sound Radiation from Vibrating Surfaces," *Jour. Acous. Soc. Am.,* 36, No. 7 (July 1964), 1305-1313.

21. Schenck, H. A. "Formulation and Solution of Acoustic Radiation Problems as Integral Equations—Selected References," Naval Undersea Warfare Center, San Diego, California, November 1967.

22. Keller, J. B. "The Geometrical Theory of Diffraction," *Calculus of Variations and Its Applications.* American Math Soc., McGraw Hill, NYC, 1958.

23. Officer, C. B. *Introduction to the Theory of Sound Transmission.* McGraw Hill, NYC, 1958.

11

the interaction of acoustic waves and elastic structures

The previous chapter discussed the effect on acoustic waves of boundaries whose behavior was prescribed, i.e., rigid boundaries, free boundaries, et cetera. The only occasion on which the material properties of the boundaries would be significant was in the penetrable case where the boundary represented an interface between two acoustic media of different properties. Because of this restriction, that discussion could concentrate on an alternate mathematical formulation to the usual wave equation, that using integral equations based on Kirchoff's time-retarded potential integral equation form or Helmholtz's time harmonic integral equation form of the wave equation, which was applicable to arbitrary geometries.

This discussion will consider an extension of the previous topic to one in which the acoustic field affects the boundaries as well as vice-versa, i.e., a true interaction problem. Apart from the penetrable case mentioned above, which can be treated by the methods considered in Chapter 10, the next class of problems which are both manageable and of interest is that involving an acoustic fluid—elastic body interaction, i.e., problems in

which the boundary or obstacle may be taken to have an elastic material behavior.

Because the physics of these problems is somewhat more complicated than that previously considered, the geometries for the most part shall be limited to very simple shapes, e.g., circular cylinders, spheres, et cetera. Since these geometries are of practical interest as well, this limitation is not particularly restrictive insofar as application of these results is concerned. However, some mention shall be made of methods of treatment of arbitrary geometries towards the end of this discussion.

The group of problems to be considered explicitly shall again be limited to exterior problems—problems involving an elastic body immersed in an infinite acoustic fluid—and shall consist primarily of transient and time harmonic radiation and scattering problems. These are quite analagous to the group of problems considered previously which may in fact be considered limiting examples of these wherein the elastic resistance to deformation of the immersed obstacle becomes either extremely large (rigid boundary) or extremely small (free boundary) relative to the compressibility of the acoustic fluid. In addition, intermediate cases may be considered; for example, the transient elastic response may lag behind the acoustic field to such an extent that these fields may be uncoupled, at least over some early range of time.

There are a number of other significant fluid-solid interaction problems which are of a different type than those mentioned above, e.g., gravity wave interaction with structures such as ships and piers, hydroelastic problems (analogous to aerodynamic flutter) and sloshing of fuel in a partially filled elastic container. These problems require a significantly different formulation than that to be described here. In addition, there are a number of acoustic-elastic problems which are physically of the type to be considered here but which represent interior problems, e.g., acoustic fluids within hollow elastic tubes (water hammer, blood flow, etc.). Although techniques to be discussed apply to these problems as well, it is felt that the methods used could be best illustrated by concentrating on one class of problems—exterior scattering and radiation problems.

Even though the physics of the problem has been complicated by the introduction of elastic effects, these shall be limited to two special cases to simplify the treatment. The obstacles are taken as either solid isotropic homogeneous elastic bodies or thin-walled isotropic homogeneous elastic shells, with emphasis on the latter as more representative of realistic structures in the ocean.

In scattering problems, the former are expected to behave more like completely rigid obstacles while the latter might be expected to behave more like free boundary obstacles. Although the equations of motion for a thin-walled elastic shell are only approximations of the general equations of motion for an elastic solid, even for the simple shapes considered here, they are generally more complicated than the corresponding equations of elasticity for a solid body. Unfortunately, the problems of greatest physical interest involve thin-walled shells rather than solid elastic bodies, except perhaps in geophysical applications. This is particularly true for submerged or floating structures where the mass of the structure—only a portion of which is represented by the surrounding elastic shell—is of the same order of magnitude as the mass of the displaced water.

Also, it is true that the presence of the surrounding fluid has a much more pronounced effect on the motion of a thin walled shell than on the motion of a solid elastic body. Therefore, one would expect the boundary coupling to be far more important for thin-walled shell problems, except for those frequencies near the natural frequencies of the elastic solid where coupling effects may also be significant in that case as well. For these reasons, only a cursory treatment will be given to those problems involving solid obstacles and the bulk of the discussion will concentrate on thin-walled elastic shells.

The methods of solution to these problems are discussed in general, attempting to describe the relevant features of the mathematical approaches and the form of approximations made without becoming overly involved in mathematical detail. Unfortunately, the ability to manipulate the mathematical details seems to be the ultimate criterion in solving a specific problem. For this reason, some of the detail must be described as well to point out the practical difficulties along with the theoretical ones.

In both cases (thin-walled elastic shells and elastic solids), the problems may be categorized as follows:

1. Transient or time harmonic
2. Radiation or scattering
3. Two or three dimensional

The problems of thin-walled elastic shells further require a specification of

4. Form of the shell equations used

and, in reducing the manipulation to a feasible problem,

5. Form of the interaction boundary condition used.

These features may, of course, be considered separately from the actual mathematical techniques to be used.

The most direct approach is based on a solution of the governing elastic shell (or solid) equations and the fluid equations coupled together by the appropriate boundary conditions. For transient problems, a Fourier transform is frequently employed to eliminate the time variable, leaving the problem of inversion,[1] however.

Another approach is to use an expansion of the significant dependent variables in terms of generalized coordinates. One such method[2] uses the coefficients in a Fourier series expansion of the dependent variables as these generalized coordinates while another[3] uses the normal modes of the uncoupled problem, i.e., elastic shell (or solid) vibrating freely in vacuo, a problem whose solution is frequently available.

While there are a number of other approaches significantly different from these that will be mentioned, the bulk of the research in this area appears to be based on one or the other of these approaches. The famous "law of invariance of difficulty," however, implies that various approaches to the same problem may shift the difficulty from one area to another, but rarely eliminate it or even reduce it substantially. Any of the various approaches used may appeal to one particular user for one reason or another, but as yet it does not appear that any one approach is vastly superior to the others. Therefore, several of the different techniques are briefly described.

This, of course, does not imply the same conclusion regarding the various types of approximations used, since a better approximation will lead usually to a better result although it may not appear to be worth the additional effort in some cases. Therefore, the physical significance of the various approximations used must be understood if the results are to be interpreted properly. These approximations for thin-walled shells fall generally into two categories—*approximation of the behavior of the elastic shell,* e.g., by various orders of shell theory[4]; and *approximation of the fluid response,* usually by some approximation of the coupling boundary condition. Though the significance of these approximations may sometimes be discussed a priori, it is usually necessary that both higher order theory solutions and experimental results be available for comparison, at least in some special cases, to produce a significant measure of confidence in the value of the results based on these approximate theories. Consequently, it may appear that the same problem has been solved repeatedly while in reality each new solution usually represents an attempt to account for some higher order approximation or some direct physical interpretation of the results.

The relevant equations of motion for an acoustic fluid, and an elastic solid are contained in the appendix. Equations describing the motion of thin-walled elastic shells require a deeper discussion and may be found in standard texts on shell dynamics[5] and in Reference 4, which is a review article on elastic shell dynamics.

Problems of acoustic fluid-elastic solid interaction can be solved by directly applying eigenfunction expansions (separation of variables) for those particular simple geometries considered here. A typical solution is the time-harmonic scattering of plane acoustic waves by infinite solid circular cylinders and solid spheres given in Reference 6. The technique of solution is very straightforward. The result of greatest interest is that if the incident wave frequency is less than that of the first symmetrical normal mode of free vibration of the solid elastic obstacle and if the density of the solid is greater than that of the fluid, there appears, on grounds of theoretical calculation and experiment, to be little difference between the acoustic-elastic solution and that for a rigid immovable obstacle immersed in an acoustic fluid.

However, significant changes in the scattering pattern and cross-section can occur at frequencies in the vicinity of the natural frequencies of vibration of the scatterer immersed in the fluid. These frequencies are generally very close to the natural frequencies of the scatterer in vacuo. This is not the case for thin-walled elastic shells. In fact, in some instances these may be treated as completely free boundaries.

Similar methods may be used in the treatment of radiation and transient scattering from solid elastic obstacles. However, the remainder of this discussion shall be spent on the more significant, and in some ways more difficult, problems in ocean engineering of thin-walled elastic shell structures.

While a detailed analysis of general shell theory is beyond the scope of the present discussion, there are some aspects of shell theory as related to circular cylinders and spheres which should be examined at this point. These concern the physical significance of some of the approximations made in the thin shell theories used in an example to be given later. One essential question involves the relative importance of extensional versus flexural effects, both of which are contained in the general theory. For closed surface shells such as the sphere, a theorem[7] states that significant deformation requires extension of the middle surface of the shell. This indicates, then, that extensional effects for these problems should be more significant than flexural effects and in turn implies that membrane theory should be satisfactory for these problems, at least over some frequency range. Membrane theory is markedly simpler than flexural theory or a combined general theory, allowing a substantial reduction in calculations.

However, the use of membrane theory in static problems leads to some difficulty near points of discontinuous loading where bending effects actually predominate. In dynamic problems, however, it has been shown[8] that the presence of inertial terms will resolve most of this difficulty and dynamic membrane theory has been used in actual problems, e.g., transient scattering of plane acoustic waves by spherical shells.[9]

For open surface shells, such as the infinitely long circular cylinder without internal reinforcement under a loading constant in the axial direction, the opposite result occurs. There, flexural effects appear to predominate, apart possibly from a uniform dilatation (or expansion) of the cross section. This assumption has been used in radiation,[10] time-harmonic scattering[2] and transient scattering[3] examples. A later treatment of the transient scattering case[11] included extensional effects with substantially more effort and little

change in results except at very early times thereby justifying to some extent the original treatment. Many problems such as those involving reinforced cylindrical shells require both extensional and flexural terms with correspondingly lengthier calculations.[12]

Another point of interest is the question of the effect of rotatory inertia and transverse shear. In beam theory, these terms may be significant in describing high frequency response; some shell theories used in the treatment of these acoustic-elastic interaction problems include these terms as well.[13]

For the actual problem to be considered here in detail however, the equations of thin-walled elastic shell theory can be hidden by use of a modal expansion employing the normal modes and natural frequencies of the same shell in free vibration as generalized coordinates.[3,11,12] This approach assumes that these free vibration modes are known from some previous calculation and uses them in the solution of the interaction problem. Thus, the only place in which the shell equations arise in this approach is in the calculation of these free vibration modes, thereby hiding these equations in previous calculations which for the simple geometries considered here have frequently (but not always) been carried out in some earlier unrelated investigation.

The most common other approach is to use the equations of motion of a thin-walled shell directly with the corresponding acoustic equations,[1] although some time-harmonic problems in radiation and scattering use coefficients of a Fourier series expansion for the shell displacements as generalized coordinates and obtain the corresponding equations of motion by a Lagrange equation.[2]

However, the modal approach seems best suited for the purpose of discussing thin elastic shell problems without using thin elastic shell equations. This approach can therefore concentrate on the interaction problem without adding the difficulties of explicitly using the thin shell equations as well.

To illustrate this approach, the problem discussed in Reference 3 will be discussed here in

Figure 11-1. Geometry for cylindrical shell-acoustic wave interaction problem.

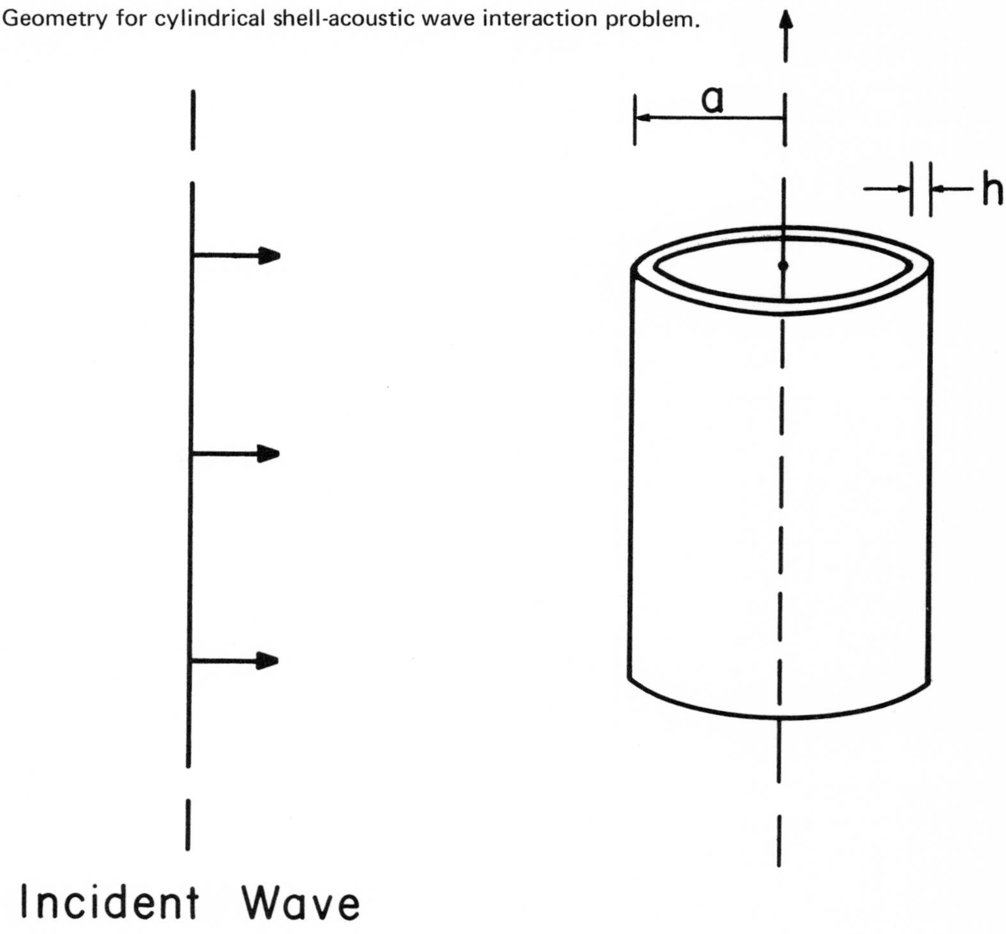

Incident Wave

detail followed by a brief description of later extensions for this case.

Considered now is the transient scattering of a plane acoustic step wave by an infinitely long thin-walled elastic circular cylinder of radius a, thickness h, Youngs Modulus E, cross sectional area A, cross sectional moment of inertia I and Poisson ratio ν with the incident wave front parallel to the cylinder axis so as to give a two-dimensional problem (Figure 11-1). The deformations in the nth mode of the shell freely vibrating in a vacuum are in general given by the inward radial

displacement, $w = \cos n\theta$ and the tangential displacement, $v = c \sin n\theta$, where c is a constant (depending on n) and the axial displacement is zero, i.e., plane strain.[6]

For n = o, there is only one natural frequency—that for a pure dilatation—while for each $n \geqslant 1$ there are two natural frequencies, one controlled primarily by the bending stiffness and one by extensional effects. As a first approximation, since the shell is an open cylinder, the extensional effects are assumed to be negligible (except for the n = o mode) compared to flexure; and the corresponding extensional frequencies shall be neglected, leaving only one frequency and mode shape for each value of n.

These modes may then be used as generalized coordinates for use in the solution of the interaction problem. This is a classical approach to vibration problems under generalized forces which in this case involve the unknown fluid reaction to the shell motion, leading to a coupled problem.

The actual radial displacement for the shell submerged in the fluid shall now be expanded into these modes

$$w(\theta,t) = \sum_{n=0}^{\infty} q_n (t) \cos n\theta \qquad (11\text{-}1)$$

and the corresponding surface pressure on the shell will be

$$Z (\theta,t) = \sum_{n=0}^{\infty} Q_n (t) \cos n\theta \qquad (11\text{-}2)$$

Then the corresponding equations of motion for the shell may be written as

$$\ddot{q}_n + \omega_n^2 q_n = Q_n / m_n \qquad (11\text{-}3)$$

where the ω_n are the natural frequencies of the shell without the surrounding fluid, i.e., "unforced" and m_n is the generalized mass given in terms of m, the mass of the shell per unit surface area.

$$m_n, \omega_n^2 =$$

$$\begin{cases} m, EA/(ma^2 (1\text{-}\nu^2)) & :n=0 \\ m (1+1/n^2), EI (1\text{-}n^2)^2 / (m_n a^4 (1\text{-}\nu^2)) & :n \geqslant 1 \end{cases}$$

$$(11\text{-}4)$$

The acoustic field may be expanded in the same form. Calling p the known incident pressure field on the cylinder surface and ϕ the corresponding scattered velocity potential field (with scattered pressure equal to $\rho\, \partial\phi/\partial t$ where ρ is the fluid density and where the scattered velocity is $\nabla\, \phi$

$$p (\theta,t) = \sum_{n=0}^{\infty} P_n (t) \cos n\theta \qquad (11\text{-}5)$$

and

$$\phi (r,\theta,t) = \sum_{n=0}^{\infty} \psi_n (r,t) \cos n\,\theta \qquad (11\text{-}6)$$

where $\nabla^2\phi = (1/c^2)\, \partial^2\phi/\partial t^2$, i.e., ϕ satisfies the wave equation with wave speed c in the acoustic fluid. Then

$$Q_n(t) = P_n (t) + \rho\, \dot{\psi}_n (a,t) \qquad (11\text{-}7)$$

represents the total generalized force in the nth mode acting on the shell.

In addition to this equation, the radial velocity of the shell must match the radial velocity of the fluid at the surface r = a. Defining $\xi (\theta, t)$ as the known incident radial velocity such that

$$\xi (\theta,t) = \sum_{n=0}^{\infty} V_n (t) \cos n\,\theta \qquad (11\text{-}8)$$

the resulting boundary velocity condition becomes

$$\dot{q}_n = V_n + \frac{\partial \psi_n(a,t)}{\partial r} \qquad (11\text{-}9)$$

Equations 11-3, 11-7 and 11-9 then form a coupled set of differential equations to be solved for q_n and ψ_n as functions of t in terms of the known incident values P_n and V_n. By use of a Laplace Transform, this leads to

$$q_n(t) = \frac{a}{2\pi i \rho c^2} \int_{-i\infty}^{+i\infty} \left\{ \frac{\overline{P}_n(s) K_n'(s) - \rho c \overline{V}_n(s) K_n(s)}{\left[\frac{m_n}{\rho a} s^2 + \frac{a\omega_n^2}{\rho c^2} \right] K_n'(s) - s K_n(s)} \right\} e^{sct/a} \, ds$$

(11-10)

where $\overline{P}_n$, $\overline{V}_n$ are the Laplace Transforms of P_n, V_n, respectively and $K_n(x)$ is the modified Bessel function of the second kind. While the equations have in principle now been solved (see Reference 1 for such a treatment using the shell equations directly), the inversion "arithmetic" is formidable, though with present computing capabilities far from impossible. For this reason, an approximation is introduced in Reference 3 which uncouples the boundary condition by assuming $K_n'(s) \doteq -K_n(s)$, which is valid for large s or equivalently

$$\frac{\partial \psi_n(a,t)}{\partial r} \doteq - \frac{1}{c} \frac{\partial \psi_n(a,t)}{\partial t}$$

giving the scattering of each mode the character of a plane wave. This assumption gives essentially a short time approximation to the general solution using the complete boundary condition. In this case, $\dot{\Psi}_n$ may be replaced by $- c [\dot{q}_n - V_n]$ in Equation 11-7 leading to

$$m_n \ddot{q}_n + \rho c \dot{q}_n + m_n \omega n^2 q_n = P_n + \rho c V_n$$

(11-11)

which may be solved for any given P_n and V_n in a relatively simple manner. In the particular case of an incident step pressure wave, P_n and V_n are given in Reference 3 along with results for displacements, pressures, velocities and accelerations for the first three modes.

Because of the assumptions made in this approach, results were felt to be valid only for small values of time, i.e., generally much less than the

order of one transit time, $2a/c$. However, at very early values of time the mode approach convergence is poor, i.e., the number of modes required for an accurate summation is large. Of course, this result is independent of the form of boundary coupling used but rather is a result of the modal approach per se. Therefore, it would be necessary to examine the approximations made in an attempt to improve the range of application of this theory to early times. Furthermore, the maximum deflections in some modes occur at values of time greater than one transit time and accuracy at later values would also be important.

One obvious step would be to retain the extensional modes in the analysis. When this was done,[11] the solutions were found to be somewhat affected at early times—particularly for light shells, i.e., those whose surface mass is less than the equivalent mass of the fluid displaced—although the extensional response appears to be heavily damped in time. However, the conclusion is reached that the simpler theory can be used to approximate the shock effects on structures except for very early times or for elements whose frequency is comparable to the frequencies of extensional vibration of the shell. Since the modal approach itself is poor at very early times, this correction does not appear to warrant the additional effort. Results comparing these two approximate theories are shown in Figure 11-2.

Another refinement[14] of the method was introduced in an attempt to improve the form of the boundary coupling condition and still avoid the extensive calculations required by using the exact boundary condition. Rather than employing the plane wave approximation, an exact solution of the wave equation was written as

$$\frac{\partial \psi_n}{\partial r} = -\frac{1}{c} \frac{\partial \psi_n}{\partial t} - \frac{1}{r} \int_0^\infty g_n \cdot F_n (ct - r\cosh u) \cdot \cosh nu \cdot du \qquad (11\text{-}12)$$

where

$$g_n = \frac{1 + n \cdot \sinh u \cdot \tanh nu}{1 + \cosh u}$$

If g_n is approximated by a constant value $\widehat{g_n}$, this result becomes

$$\frac{\partial \psi_n}{\partial r} = -\frac{1}{c} \frac{\partial \psi_n}{\partial t} - \frac{\widehat{g_n}}{r} \psi_n \qquad (11\text{-}13)$$

$\widehat{g_n}$ may be bounded as $t \to 0, \infty$ by

$$0 \leqslant g_0 \leqslant \tfrac{1}{2} : n = 0$$

$$\tfrac{1}{2} \leqslant g_n \leqslant n : n \geqslant 1$$

A mean value of $\widehat{g_n}$ is introduced into the calculations which now involve a third order dif-

ferential equation, using $T = ct/a$, $\lambda_n = \dfrac{m_n}{\rho a}$ and $\sigma_n = \lambda_n \cdot \omega_n^2 (a/c)^2$

$$\lambda_n \frac{d^3 q_n}{dT^3} + (\lambda_n \overline{g}_n + 1) \frac{d^2 q_n}{dT^2} +$$

$$\sigma_n \frac{dq_n}{dT} + \sigma_n \overline{g}_n q_n =$$

$$\frac{a}{\rho c^2} \left\{ \frac{dP_n}{dT} + \overline{g}_n P_n + \rho c \frac{dU_n}{dT} \right\} \qquad (11\text{-}14)$$

which is almost as easy to solve as Equation 11-11, but which would be expected to be valid over a longer range in time, since the cylindrical character of the wave has been kept. A comparison of the solutions based on the plane wave approximation and the cylindrical wave approximation with the exact solution for the mode $n = 0$ is shown in Figure 11-3. Two later papers have treated other

Figure 11-2. Comparison of inextensional theory to a theory including extensional effects (after Baron[11]).

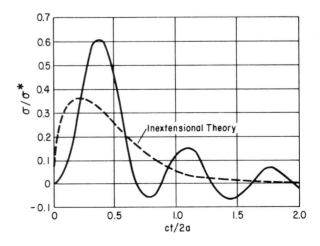

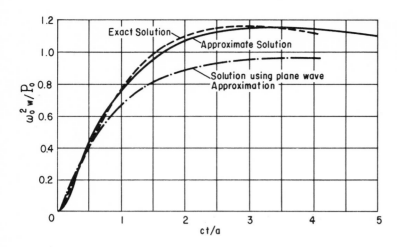

Figure 11-3. Comparison of cylindrical wave approximation to plane wave approximation (after Haywood[14]).

aspects of this same problem. One considered the numerical far field solutions for the limiting cases of a rigid fixed cylinder and a rigid floating cylinder by means of the same modal approach.[15] The other used a completely different technique, where the solution domain is considered as a folded Riemann sheet.[16] This same technique has been used successfully in acoustic scattering problems as a "creeping wave" method. The shell equations in this case were introduced directly and included not only extension as well as flexure, but also rotatory inertia and shear deformation. This approach is particularly well suited for those short time intervals where the modal approach requires a large number of terms.

In a brief summary of the treatment of this particular problem, it would appear that very early time solutions (much less than one transit time) for which the plane wave approximation inextensional theory holds requires many modes and, having neglected higher order extensional effects, may not be applicable even if all of the significant modes were included.

For solutions of the order of one transit time, $t \sim 2a/c$, the cylindrical wave approximation appears to be better than the plane wave approximation, and convergence within a reasonable number of modes would be expected. This range of time is of interest, since it is of the order of the period of the fundamental mode of thin steel

shells in water. To illustrate these results, the specific example of a circular steel cylinder submerged in sea water is considered, taking a radius, a = 25 ft, a thickness, h = .286 ft such that a/h = 87.33. The density of sea water is 64 lb /ft^3 and steel is 490 lb /ft^3, giving ρ water/ρ shell $\simeq 0.13$, while Young's modulus is taken as 30x10^6 psi and Poisson's ratio as 0.3. The speed of sound in sea water is approximately c = 5000 ft/sec.

Then the transit time of the incident wave over the cylinder is $t_o = 2a/c = 10^{-2}$ sec, which will serve as a time scale for the problem.

The cross-sectional area of the shell is

$$A = 2\pi\, ah = 44.8 \text{ ft}^2$$

while the moment of inertia is

$$I = 2\pi\, a \cdot \frac{1}{12}\, h^3 = 0.305 \text{ ft}^4$$

and the mass of the shell per unit circumference per unit length is

$$m = \rho_s \cdot h \cdot 1 \text{ foot} = 140. \text{ lb (mass) / ft.}$$

Then the dilatational natural frequency (n =o) is $\omega_o = 8.85 \times 10^3$ rad/sec with a corresponding period of 7.07×10^{-4} sec, which is much shorter than a transit time. The n = 1 "flexural" mode is really a translational mode with no free vibration, and the first true flexural natural frequency is n = 2 with $\omega_2 = 3.50 \times 10^1$ rad/sec with a period of 1.79×10^{-1} sec, which is roughly six transit times.

The corresponding higher extensional modes are given by

$$\overline{\omega}_n \simeq (1 + n^2)^{1/2} \omega_0$$

and are comparable in magnitude to ω_0, at least for the first few modes. The higher flexural modes are given by

$$\omega_n \simeq \sqrt{\frac{I}{a^2 A}} \cdot n \cdot \left(\frac{n-1}{n+1} \right) \omega_0$$

and are comparable in magnitude to ω_2, again for the first few modes.

It would therefore seem that very early time behavior is strongly influenced by extensional behavior. However, these modes are damped substantially by the end of the first transit time by the presence of the surrounding fluid, as may be seen from Figure 11-2.

The next simple geometry that can be considered which retains the same number of independent variables is the sphere. The problem of a thin-walled elastic spherical shell scattering an incident step wave was solved[9] using membrane theory, since the sphere is a closed surface as opposed to the open circular cylinders discussed above. This simplifies the equations of motion considerably, and results were obtained by a direct eigenfunction expansion without approximations of the boundary coupling. The resulting series gave reasonable convergence for the hoop stresses in the shell at early times. However, the radial acceleration and fluid pressure gave much poorer convergence due probably to the discontinuity in pressure of the incident wave and the neglect of bending effects which might be expected to be significant at this location, although a later investigation[8] of bending effects in dynamic response of thin shells to moving discontinuous loads does indicate that there are many circumstances where membrane theory is adequate.

In extending the geometry from two to three-dimensional problems, the case of greatest interest is that of uniform circular cylindrical shells with periodic reinforcement in the axial direction. If the supports are considered rigid, the effect of extension of the surface of a cylinder segment between supports becomes predominant over flexural effects but not to the point where the latter can be completely neglected.

Considering the problem of time-harmonic radiation for this geometry under periodic axial loading as the simplest example, both the modal[12,17] and the Lagrange equation formulations[18] found it necessary to include flexural effects in some approximate manner along with the predominantly extensional behavior. The frequencies of free vibration of the submerged cylinder were calculated[17,19] as well as the cylinder response to a forced motion. Two interesting conclusions arose during these studies. For a forced motion frequency, Ω, less than $\pi c/L$ where c is the acoustic sound speed and L the half wave length in the axial direction, there is no energy radiated from the shell into the fluid, since the pressure and the shell displacement will be in phase with the applied load. For a steel shell in water with L=10.0 ft, this cut-off frequency is about 1600 rad/sec or 250 cps. The other point is that resonant frequencies exist where the magnitude of the pressure becomes very large, although not infinite in any practical situation where structural damping would have to be included. A useful discussion of these points is given in Reference 20.

The corresponding time-harmonic scattering problems for either uniform circular cylinders[2] or reinforced circular cylinders[21] have again been solved in a similar straightforward manner and shall not be discussed here. However, recent studies have been made of time-harmonic acoustic scattering by solid elastic circular cylinders[22] and by elastic circular cylindrical shells[23] by means of a creeping wave analysis which examines the surface waves generated at the fluid-shell interface. This approach allows for a physical interpretation of the wave propagation and scattering in terms of geometrical acoustics. This leaves open the possi-

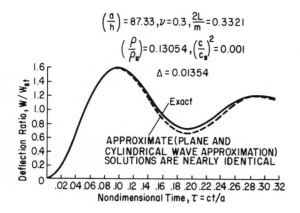

$$\left(\frac{a}{h}\right)=87.33, \nu=0.3, \frac{2L}{m}=0.3321$$

$$\left(\frac{\rho}{\rho_s}\right)=0.13054, \left(\frac{c}{c_s}\right)^2=0.001$$

$$\Delta = 0.01354$$

Exact

APPROXIMATE(PLANE AND CYLINDRICAL WAVE APPROXIMATION) SOLUTIONS ARE NEARLY IDENTICAL

Deflection Ratio, W/W_{st}

Nondimensional Time, $\tau = ct/a$

Figure 11-4. Exact solution of Reference 24 compared to approximate solutions (after Herman and Klosner[24]).

$$\left(\frac{a}{h}\right)=87.33, \ \nu=0.3, \ \frac{2L}{m}=0.3321$$

$$\left(\frac{\rho}{\rho_s}\right)=0.13054, \left(\frac{c}{c_s}\right)^2=0.001, \ \Delta=0.9828$$

E = Exact Solution
C_1 = Cylindrical Approximation, $\overline{S}_o$ = 0.5
C_2 = Cylindrical Approximation, S_o = 0.363
P = Plane Wave Approximation

Deflection Ratio, W/W_{st}

E
C_1
C_2
P

Nondimensional Time, $\tau = ct/a$

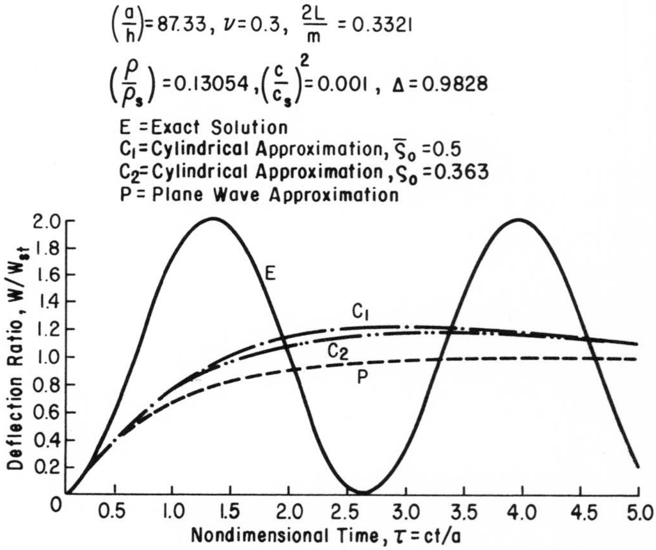

Figure 11-5. Exact solution of Reference 24 compared to approximate solutions (after Herman and Klosner[24]).

bility of an extension of this approach to more complicated geometries.

Returning once again to transient problems, a simple problem[24] has been solved exactly which may be used as a test case for the plane wave and cylindrical wave approximations mentioned earlier. The problem concerns the transient response of an infinitely long, periodically simply supported, circular cylindrical shell subjected to a sudden circumferentially uniform pressure distribution, which is also periodic along the axis.

The resulting shell equation is quite simple because of the symmetries in the problem, and the

exact solution is obtained as a Fourier transform inverse which may be evaluated numerically. Results showed a non-radiating and a radiating contribution. Since the plane wave and cylindrical wave approximations did not contain any non-radiating term, it would be expected that these approximations would be poor when the non-radiating term was significant. Typical results are given in Figures 11-4 and 11-5. A later paper[25] solves the problem of transient scattering of a plane acoustic step wave by a reinforced circular cylinder using an "acceleration and velocity approximation," which replaces the coupling boundary condition by one

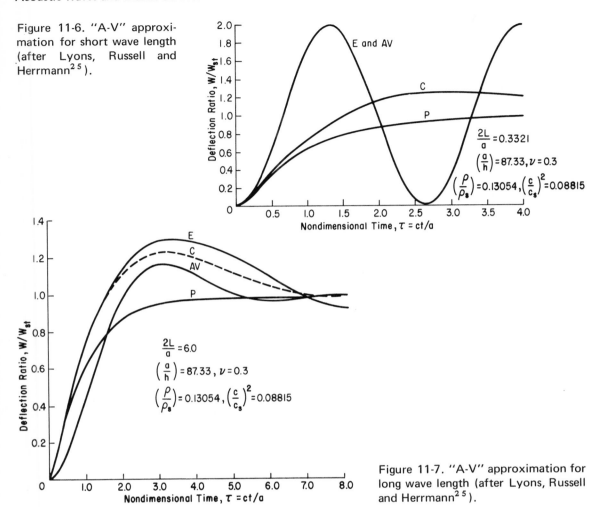

Figure 11-6. "A-V" approximation for short wave length (after Lyons, Russell and Herrmann[25]).

$$\frac{2L}{a} = 0.3321$$
$$\left(\frac{a}{h}\right) = 87.33, \nu = 0.3$$
$$\left(\frac{\rho}{\rho_s}\right) = 0.13054, \left(\frac{c}{c_s}\right)^2 = 0.08815$$

$$\frac{2L}{a} = 6.0$$
$$\left(\frac{a}{h}\right) = 87.33, \nu = 0.3$$
$$\left(\frac{\rho}{\rho_s}\right) = 0.13054, \left(\frac{c}{c_s}\right)^2 = 0.08815$$

Figure 11-7. "A-V" approximation for long wave length (after Lyons, Russell and Herrmann[25]).

in which the surrounding acoustic fluid is replaced by a "virtual mass" term and a "damping" term corresponding to the reaction which the fluid would have on the shell were the shell vibrating at only one frequency chosen to be the natural frequency of the shell in an incompressible fluid. In this paper the problem discussed in Reference 24 was also solved by this "A-V" approach, and a comparison is shown in Figures 11-6 and 11-7. The primary difference between the two "A-V" cases shown is that the poorer result obtained in Figure 11-7 corresponds to a much larger reinforcement spacing.

Finally, a modified cylindrical wave approximation[26] has been suggested for reinforced shell problems which accounts for axial changes. This approach is valid in all cases for short time and in many cases for long time. There are some cases, however, where the solution is unstable after a while.

While the previous discussions were all concerned with separable geometries, a reasonable question arises as to what to do if the geometry is not one of these. One possible approach for solid elastic bodies is an adaptation of the integral equation formulation described in Chapter 10.

While this approach has been extended to time-harmonic[27] and transient[28] elastic wave scattering, it appears to have been applied to only one interaction problem of this type.[35] For thin shell problems, there are some other techniques available, however. One uses a simple source integral equation approach,[29] which replaces the elastic structure by "rigid piston" elements which are related to the in vacuo vibrations of the shell and assumed to be found previously, in much the same manner as Reference 12.

Another approach was carried out for transient scattering problems[30] in which the major structural effects are produced by bending terms which have periods many times the transit time of the incident wave across the shell. A typical example of this would be a hemisphere or hemicylinder floating at a free surface. In this case the n=o mode (dilatation) is not present, and the effect of the elastic response of the shell would not affect the velocity distribution appreciably at the end of the first transit time. These velocities could then be calculated for a scattering obstacle having the same geometry but consisting of unconnected mass elements. After one or two transit times, these velocities are essentially constant and can then be used to determine the corresponding bending stresses and deformations of the shell as an uncoupled initial velocity problem.

Results in this case and a corresponding case for a floating cylinder with a rectangular cross section further indicate that the shell may be considered as a free surface (i.e., no mass) for those problems in which the mass per unit surface area of the shell was small compared to the mass per unit volume of the fluid times an appropriate length scale of the problem.

This approach has been used to determine the rigid-plastic response of a floating cylinder of rectangular cross section[31] using the free boundary condition solution[32] obtained earlier by techniques described in Chapter 10.

In conclusion, a brief review of some aspects of acoustic fluid-elastic solid dynamic interaction problems has been given with particular emphasis on the dynamic response of submerged thin-walled elastic shells to transient or time-harmonic acoustic loading. A recent review of this area concerned primarily with the effect of loads traveling parallel to the axis of the submerged body is given in Reference 33. These problems are clearly of importance in terms of questions that range from detecting and identifying submerged obstacles (low intensity acoustic loading) to protection (or destruction) of submerged obstacles (high intensity loading). It is equally clear that this represents a current field of active research interest and that much work remains to be done. (See Reference 34.)

Appendix

Equations governing the behavior of the acoustic fluid can be conveniently represented in terms of a velocity potential ϕ such that

$$\Box_o^2 \phi = \nabla^2 \phi - \frac{1}{c_o^2} \frac{\partial^2 \phi}{\partial t^2} = F_o (\bar{r}, t)$$

where F_o is some forcing function and $c_o = (dp/d\rho)^{1/2}$ represents the speed of propagation of acoustic disturbances. The velocity field $\bar{v}$ and excess pressure ρ are then

$$\bar{v} = -\nabla \phi$$

$$p = \rho_o \, \partial\phi/\partial t$$

where ρ_o is the density of the acoustic fluid.

The equation governing the dynamic response of an isotropic homogeneous linear elastic solid is Navier's equation

$$(\lambda + \mu) \nabla (\nabla \cdot \bar{u}) + \mu \nabla^2 \bar{u} = \rho_e \, \partial^2 \bar{u}/\partial t^2$$

where $\bar{u}$ is the displacement of the elastic body whose density is ρ_e and whose material behavior is described by the Lame constants λ and μ.

By introducing displacement potentials Φ and $\overline{\Psi}$ such that $\overline{u} = \nabla \Phi + \nabla \times \overline{\Psi}$ with the gauge condition

$$\nabla \cdot \overline{\Psi} = 0$$

Navier's equation can be decomposed into two wave equations

$$\Box_D^2 \; \Phi =$$

$$\nabla^2 \Phi - \frac{1}{c_D^2} \; \partial^2\Phi/\partial t^2 = F_D \; (\bar{r}, t)$$

and

$$\Box_R^2 \; \overline{\Psi} =$$

$$\nabla^2 \overline{\Psi} - \frac{1}{c_R^2} \; \partial^2\overline{\Psi} / \partial t^2 = \overline{F}_R \; (\bar{r}, t)$$

where F_D, F_R are forcing functions and

$$c_D = [(\lambda + 2\mu) / \rho]^{1/2}$$

$$c_R = [\mu/\rho]^{1/2}$$

The corresponding strain field is given in terms of cartesian tensor components as

$$\epsilon_{ij} = \tfrac{1}{2} \, [u_{i,j} + u_{j,i}]$$

and the stress field by

$$\sigma_{ij} = \lambda \, \epsilon_{kk} \, \delta_{ij} + 2\mu \, \epsilon_{ij}$$

List of Symbols

a   Cylinder radius
A   Cross-sectional area
c   Speed of sound in acoustic fluid
E   Young's modulus
g_n Function introduced in Equation 11-12
h   Shell thickness
I   Cross-sectional moment of inertia

K_n Modified Bessel function of second kind of order n
L   Half wave length in axial direction
m   Mass of shell per unit surface area
m_n Generalized mass (Equation 11-4)
n   Modal index
p   Incident pressure field
P_n Modal incident pressure (Equation 11-5)
q_n Modal radial displacement (Equation 11-1)
Q_n Modal surface pressure (Equation 11-2)
r   Polar coordinate
s   Laplace transform variable
t   Time
T   Non-dimensional time = $t \cdot (c/2a)$
v   Tangential displacement
V_n Modal incident radial velocity (Equation 11-8)
w   (Inward) radial displacement
z Surface pressure
σ_n Non-dimensional frequency
Ω Forcing frequency
λ_n Non-dimensional generalized mass
ξ Incident radial velocity
p   Fluid density
ϕ Scattered velocity potential
Ψ_n Modal scattered potential (Equation 11-6)
v   Poisson ratio
o   Polar coordinate
w_n Natural frequency (Equation 11-4)

References

1. Carrier, G. F. "The Interaction of an Acoustic Wave and an Elastic Cylindrical Shell," Cont. N70NR-35810, Tech. Rep. 4, Brown Univ., R. I., October 1951.

2. Junger, M. C. "Sound Scattering by Thin Elastic Shells," *Jour. Acous. Soc. Am.* 24, No. 4 (July 1952), 366-373.

3. Mindlin, R. D. and H. H. Bleich. "Response of an Elastic Cylindrical Shell to a Transverse Step Shock Wave," *Jour. App. Mech.,* 19, No. 2 (June 1952), 189-195.

4. Kalnins, A. "Dynamic Problems of Elastic Shells," *App. Mech. Rev.,* 18, No. 11 (November 1965), 867-872.

5. Goldenveizer, A. "Theory of Thin Elastic Shells," Pergamon Press, NYC, 1961.

6. Faran, J. J. Jr., "Sound Scattering by Solid Cylinders and Spheres," *Jour. Acous. Soc. Am.,* 23, No. 4 (July 1951), 405-418.

7. Jellet, J. H. "On the Properties of Extensible Surfaces," *Roy Irish Acad Trans,* 22, Part V (1855).

8. Mann-Nachbar, P. "On the Role of Bending in the Dynamic Response of Thin Shells to Moving Discontinuous Loads," *Jour. Aero. Sci.* 29, No. 6 (June 1962), 648-657.

9. _____. "The Interaction of an Acoustic Wave and an Elastic Spherical Shell," *Quar App Math,* 15, No. 1 (1957), 83-93.

10. Junger, M. C. "Vibrations of an Elastic Shell in a Fluid Medium and the Associated Radiation of Sound," *Jour App Mech,* 19, No. 4 (December 1952), 439-445.

11. Baron, M. L. "The Response of a Cylindrical Shell to a Transverse Shock Wave," *2nd U.S. National Congress of App. Mech.,* (1954), 201-212.

12. Bleich, H. H. and M. L. Baron. "Free and Forced Vibrations of an Infinitely Long Cylindrical Shell in an Infinite Acoustic Medium," *Jour App Mech,* 21, No. 2 (June 1954), 167-177.

13. Forrestal, M. J. and G. Herrmann. "Response of a Submerged Cylindrical Shell to an Axially Propagating Step Wave," *Jour App Mech,* 32, No. 4 (December 1965), 788-792.

14. Haywood, J. H. "Response of An Elastic Cylindrical Shell to a Pressure Pulse," *Quart Jour Mech and App Math,* 11, No. 2 (1958), 129-141.

15. Skalak, R. and M. B. Friedman. "Reflection of an Acoustic Step Wave from an Elastic Cylinder," *Jour App Mech.,* 25, No. 1 (March 1958), 103-108.

16. Payton, R. G. "Transient Interaction of an Acoustic Wave with a Circular Cylindrical Elastic Shell," *Jour Acoust Soc Am,* 32, No. 6 (June 1960), 722-729.

17. Bleich, H. H. "Sound Radiation from an Elastic Cylindrical Shell Submerged in an Infinite Medium," *2nd U. S. National Congress of Applied Mechanics* (1954), 213-223.

18. Junger, M. C. "Dynamic Behavior of Reinforced Cylindrical Shells in a Vacuum and in a Fluid," *Jour App Mech,* 21, No. 1 (March 1954), 35-41.

19. Baron, M. L. and H. H. Bleich. "Tables for Frequencies and Modes of Free Vibrations of Infinitely Long Thin Cylindrical Shells," *Jour App Mech,* 21, No. 2 (June 1954), 178-184.

20. Bleich, H. H. "Dynamic Interaction Between Structures and Fluids," *Proc First Symposium on Naval Structural Mechanics,* Pergamon Press, NYC, 1960, 263-284.

21. Junger, M. C. "Concept of Radiation Scattering and Its Application to Reinforced Cylindrical Shells," *Jour Acous Soc Am,* 25, No. 5 (September 1953), 899-903.

22. Doolittle, R. D., H. Uberall and P. Ugincius. "Sound Scattering by Elastic Cylinders," *Jour Acous Soc Am,* 43, No. 1 (January 1968), 1-14.

23. Ugincius, P. and H. Uberall. "Creeping Wave Analysis of Acoustic Scattering by Elastic Cylindrical Shells," *Jour Acous Soc Am,* 43, No. 5 (May 1968), 1025-1035.

24. Herman, H. and J. Klosner. "Transient Response of a Periodically Supported Cylindrical Shell Immersed in a Fluid Medium," *Jour App Mech.,* 32, No. 3 (September 1965), 562-568.

25. Lyons, W. C., J. E. Russell and G. Herrmann. "Dynamics of Submerged Reinforced Cylindrical Shells," *Am Soc Civil Eng. Jour. Eng Mech Div,* 94, No. EM2 (April 1968), 397-420.

26. Russell, J. E. and G. Herrmann. "A Modified Cylindrical Wave Approximation," *Jour App Mech.,* 35, No. 4 (December 1968), 819-822.

27. Banaugh, R. "Scattering of Elastic Waves by Surfaces of Arbitrary Shape," UCRL Rep. 6779, Lawrence Radiation Laboratory, Univ. of Cal., Livermore, February 1962.

28. Shaw, R. P. "Scattering of Elastic Waves by Rigid Obstacles of Arbitrary Shape" *Jour Acous Soc. Am,* 44, No. 3 (September 1968), 745-748.

29. Chen, L. H. and D. G. Schweikert. "Sound Radiation from an Arbitrary Body," *Jour Acous Soc Am,* 35, No. 10 (October 1963), 1626-1632.

30. Baron, M. L. and H. H. Bleich. "Initial Velocity in Shells at a Free Surface Due to a Plane Acoustic Shock Wave," CU-19-56-ONR-266 (08) CE, Rep 19, Columbia Univ. Civil Eng. Dept, NY, 1956.

31. Testa, R. and H. H. Bleich. "Dynamic Response of a Floating Rigid Plastic Box," *Am Soc Civil Eng. Jour Eng. Mech Div,* 92, No. EM2 (April 1966), 131-151.

32. Shaw, R. P. and M. B. Friedman. "Diffraction of Acoustic Pulses by a Free Cylindrical Obstacle at a Free Surface," *4th U.S. NAT Congress of App Mech,* 1962, 371-379.

33. Herrmann, G. and J. E. Russell. "Forced Motions of Shells and Plates Surrounded by an Acoustic Fluid," *Proc. Symposium on the Theory of Shells to Honor Lloyd Hamilton Donnell,* D. Muster (ed.), University of Houston, 1967, 313-339.

34. Greenspon, J. E. (ed.). *Fluid Solid Interaction.,* Symposium at ASME Winter Annual Meeting, N. Y., 1967.

35. Baron, M. L., H. H. Bleich and A. T. Matthews. "Forced Vibrations of an Elastic Circular Cylindrical Body of Finite Length Submerged in an Acoustic Fluid," *Int. Jour. Solids and Structures,* 1, No. 1 (February 1965), 3-22.

12

methods of solution for water wave scattering problems

This discussion is limited to a few particular aspects of water wave scattering, which may be of interest in the solution of specific problems, with particular emphasis on those methods of solution which have only recently become feasible through advances in computer technology. It will, however, contain some unpublished and untested original material along with the usual discussion of methods currently in use and the specific problems to which these methods have been applied. A few selected references concerned directly with the various methods mentioned will be given. For a far more comprehensive view of the entire subject of water waves, the reader is referred to the classical treatments of this area.[1,2,3]

Problems considered here all involve a liquid with a free surface at which some pressure distribution is given, usually as zero. A variation of such problems involves the specification of an interface between a floating object and the fluid where the pressure distribution is not known a priori but is determined as part of the solution together with the motion of the floating object[4] but such problems are beyond the realm of this brief treatment.

The fluid is considered to be homogeneous, although some very interesting "internal gravity wave" problems[5] may arise if this assumption is not valid—these, too, are beyond the range of immediate interest. No discussion will be made of the origin of the waves which arise in these problems. It will be assumed merely that in some manner a train of periodic waves or a particular initial configuration for a transient wave will be available to serve as a forcing function appropriate to the particular problem at hand. Finally, these problems shall be taken as deterministic, i.e., a specified input producing a specific result, rather than stochastic although the latter is undoubtedly more representative of oceanographic behavior.[6]

While it may appear at this point that the range of problems has been considerably narrowed, unfortunately many more restricting assumptions shall be required before a reasonable treatment is possible. These shall be developed during the appropriate derivations of the governing equations. However, one last word of introduction seems appropriate. The field of water wave theory appears to be equally divided between practicing engineers who require simple solutions to complex problems and applied mathematicians who developed complicated solutions for "simple" problems—e.g., investigate questions of existence, uniqueness, et cetera. While there is clearly a need for both views, it is hoped that·this present treatment will be of use in describing some intermediate level approaches to intermediate level problems, i.e., problems involving complicated geometries (e.g., other than planar or circular) using theories which are general enough to include such aspects as nonlinear behavior and viscous effects.

Problems to be considered may be separated into various categories, as was the case in Chapter 10. They may be either time harmonic, corresponding to an infinitely long train of periodic waves or transient, corresponding—for example—to a solitary wave. They may be exterior problems where an incident wave which has been generated in some unspecified manner impinges on an obstacle i.e., scattering, or interior problems, such as resonance of a closed basin or tank sloshing. Because of the free surface, there may be a further distinction as to where boundaries are located. In a scattering problem, for example, the obstacle may or may not extend through the upper surface and may or may not extend to the bottom. While the vast majority of water wave problems treated are of the time-harmonic exterior scattering type, many of the solution techniques to be discussed will be equally valid for other types of problems as well, e.g., transient waves.

There are, unfortunately, a large number of approximate formulations available for the discussion of water wave problems. The discussion at this point will be concerned with the development of two of the more commonly used forms.

The first theory is for irrotational motion of an inviscid and incompressible fluid without surface tension.[2] The assumption of zero viscosity implies that the fluid will remain irrotational if at some instant it was irrotational. Therefore, the assumption of irrotationality is implicit if there is an irrotational state at any instant, e.g., if the motion starts from rest under the action of conservative forces.

Incompressibility implies that the continuity of mass equation can be written as

$$\nabla \cdot \overline{U} = 0 \qquad (12\text{-}1)$$

where $\overline{U}$ is the velocity of a particle of fluid and is measured in Eulerian coordinates. In a Cartesian coordinate system, $\overline{U}$ is written as (u, v, w). Irrotationality implies the existence of a velocity potential ϕ such that

$$\overline{U} = \nabla \phi \qquad (12\text{-}2)$$

or in combination with Equation 12-1

$$\nabla^2 \phi = 0 \qquad (12\text{-}3)$$

The lack of viscosity implies that the stress field tensor $\overline{\overline{\tau}}$ may be represented in terms of a pressure p

$$\overline{\overline{\tau}} = -p \cdot I \qquad (12\text{-}4)$$

where I is the unit tensor, thereby implying that no shear stresses exist and that all normal stresses at a point are equal to the same value, -p. The equation of motion then implies

$$\nabla \cdot \overline{\overline{\tau}} - pg\hat{k} = \rho \quad D\overline{U}/Dt \qquad (12\text{-}5)$$

where D/Dt represents the material derivative, i.e.,

$$D\overline{F}/Dt = \partial F/\partial t + (\overline{U} \cdot \nabla)\overline{F} \qquad (12\text{-}6)$$

and the coordinate axes are chosen so as to have z positive upward. The equation of motion can be written in terms of the velocity potential

$$-(\nabla p)/\rho - g\hat{k} = \nabla \cdot (\partial\phi/\partial t + \tfrac{1}{2}\nabla\phi\cdot\nabla\phi) \qquad (12\text{-}7)$$

which may be integrated to give

$$p/\rho + gz + \partial\phi/\partial t + \tfrac{1}{2}(\nabla\phi)^2 = C(t) \qquad (12\text{-}8)$$

where C is an arbitrary function of time which may be absorbed into the definition of ϕ without changing any of the other equations involving ϕ i.e., Equation 12-8 may be written with C(t) zero.

Equations 12-3 and 12-8 are valid everywhere in the domain of the fluid and may be solved to give the velocity potential and then the pressure, respectively. However, boundary and initial conditions are required to completely prescribe the problem. The "initial" conditions may specify some behavior at a specific value of time (e.g., t = 0) or, in time-harmonic problems, may prescribe some time-harmonic incident wave field plus some restriction on the solution as the independent variables become infinitely large, i.e., a radiation condition. The boundary conditions may be given on a fixed surface; for example, the normal velocity $U_n = \partial \phi / \partial n$ is zero on a rigid boundary, while the pressure (above hydrostatic) may be taken as zero on a pressure release or "free" boundary (but one fixed in space). This latter condition is different from the free surface (fluid-air interface) boundary condition, which is applied at a surface whose location is unknown, i.e., which moves. If the undisturbed fluid level is taken as z = 0, the disturbed surface may be described by

$$z = \zeta(x, y, t) \qquad (12\text{-}9)$$

Since a particle of fluid on the free surface will remain on the surface, there is a geometrical constraint on ζ

$$D(\zeta - z)/Dt = \partial \zeta / \partial t + \nabla \phi \cdot \nabla (\zeta - z) = 0$$
$$(12\text{-}10)$$

In addition, there is a mechanical boundary condition in that the pressure field on this surface is given as p_A.

$$\partial \phi / \partial t + \frac{1}{2} (\nabla \phi)^2 + g\zeta = -p_A/\rho \qquad (12\text{-}11)$$

Equations 12-10 and 12-11 are evaluated at z = ζ i.e., at the unknown surface.

Therefore, although the governing differential equation Equation 12-3 on ϕ is linear (i.e., Laplace's equation), the free surface boundary conditions (Equations 12-10 and 12-11) are nonlinear, resulting in an overall nonlinear problem.

For many physical problems, the amplitude of the disturbances from a condition of rest are small

enough to allow the nonlinear terms to be neglected. When this is done, the boundary conditions in Equations 12-10 and 12-11 become

$$\partial \zeta / \partial t - \partial \phi / \partial z = 0 \qquad (12\text{-}12)$$

and

$$\partial \phi / \partial t + g\zeta = 0 \qquad (12\text{-}13)$$

where p_A has been taken for convenience to be zero and where these conditions are applied at the surface z = 0 rather than z = ζ to be consistent with the order of approproximation made.[2] These equations may be combined to give a single condition on ϕ alone

$$\partial^2 \phi / \partial t^2 + g \, \partial \phi / \partial z = 0 \qquad (12\text{-}14)$$

At this point, a question may be raised as to the conditions under which such a linearization may be valid. If the particular case of a time-harmonic, linear, progressive wave is considered, the neglected terms are of the order (wave amplitude/wave length) less than those terms kept. It may be concluded that the linear theory should be valid whenever this quantity is small compared to one. Unfortunately, nonlinear effects may sometimes accumulate, and an additional useful restriction is that the longitudinal distance which the wave travels is not extremely long, i.e., very many wave lengths. In addition, there are many phenomena of interest, such as the breaking of waves, which linear theory cannot reproduce. However, the theory of small amplitude waves has been found generally useful and forms the basis of much of both the mathematical and the practical interest in this field (see Figure 12-1*).

A second theory frequently encountered is that of long wave-shallow water waves in which the representative wave length of the problem is assumed to be large compared to the depth of the water h(x,y). Practical problems of interest involving this

*Figure 12-1 was taken from Reference 30, which appeared as this chapter was being prepared.

assumption include the theory of tidal motion, tsunamis, et cetera. This theory is obtained from the original equations of motion and does not employ a velocity potential, since it is not an irrotational flow theory.[2]

The underlying assumption in the classical nonlinear shallow water theory is that the pressure distribution is taken to be purely hydrostatic, i.e.,

$$(p) / \rho = -g (z - \zeta) \qquad (12\text{-}15)$$

which implies that vertical acceleration is neglected and that p_A has again been taken zero. Then, derivatives of p in the x and y directions are independent of z, implying that horizontal accelerations are also independent of z, which implies that horizontal velocities are independent of z at time t if they were so at some previous time. Then, the original boundary conditions on the free surface may be written conveniently in terms of velocities—i.e., from equations 12-5 and 12-10 on z = ζ,

$$\partial u / \partial t + u \, \partial u / \partial x + v \, \partial u / \partial y =$$

$$-g \cdot \partial \zeta / \partial x = -\frac{1}{\rho} \frac{\partial p}{\partial x} \qquad (12\text{-}16)$$

$$\partial v / \partial t + u \, \partial v / \partial x + v \, \partial v / \partial y =$$

$$-g \cdot \partial \zeta / \partial y = -\frac{1}{\rho} \frac{\partial p}{\partial y} \qquad (12\text{-}17)$$

$$\partial \zeta / \partial t + \partial \left[u (\zeta + h) \right] / \partial x +$$

$$\partial \left[v \cdot (\zeta + h) \right] / \partial y = 0 \qquad (12\text{-}18)$$

which then describe the surface motion. This theory actually represents only first order approximations to a shallow water expansion theory for which many higher order approximations have been developed.[2] However, for the purposes of this discussion, only this theory and its linearized form, i.e., on z = 0

$$\partial^2 \zeta / \partial t^2 \; - \; \partial (gh \, \partial \zeta / \partial x) / \partial x \; -$$

$$\partial (gh \, \partial \zeta / \partial y) / \partial y = 0 \qquad (12\text{-}19)$$

which is obtained by combining Equations 12-16,

12-17 and 12-18, neglecting the nonlinear terms, shall be considered.

Clearly, linearized problems with constant depth require

$$\partial^2 \zeta / \partial t^2 - g h \left(\partial^2 \zeta / \partial x^2 + \partial^2 \zeta / \partial y^2 \right) = 0$$
$$(12\text{-}20)$$

i.e., a two-dimensional wave equation. Once ζ is known, u and v can be calculated from the linearized forms of Equations 12-16 and 12-17 while w follows from conservation of mass, Equation 12-1. Again, the question of the range of validity of this shallow water-long wave theory arises. Clearly, it will be best suited for those problems in which h (depth of fluid) is small compared to some representative length such as the wave length and, in the linear case, also requires small amplitudes ζ compared to h and the fluid velocities and accelerations are small enough so that the nonlinear terms in Equations 12-16, 12-17 and 12-18 may be dropped. Unfortunately, there does not appear to be any general agreement on which particular approximate theory (including several not mentioned here) applies to any specific problem. Solutions obtained in Reference 7 and compared to model experiments indicate that a simple approximate theory such as shallow water-long wave theory is apparently as good as more complicated "higher order" theories.

It has been recognized for some time that some aspects of water wave theory have an analogy to gas dynamic problems and, in particular, for the linear case to acoustic problems. Since Chapter 10 has dealt with some aspects of acoustic scattering, it would seem appropriate here to develop this analogy to make use of that material.

Considering first the small amplitude theory for time-harmonic problems involving a constant depth rigid bottom z = -h, the velocity potential φ may be taken in the form

$$\phi (x, y, z, t) = \Phi (x, y) \cdot \mathcal{Z} (z) \cdot e^{i \omega t} \qquad (12\text{-}21)$$

In order that φ satisfy Laplace's equation Equation 12-3 requires

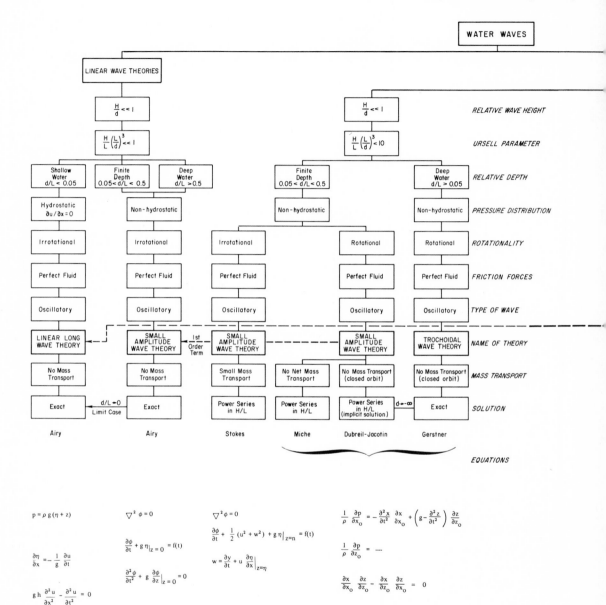

Figure 12-1. Linear, nonlinear wave theories (after LeMéhauté[30]).

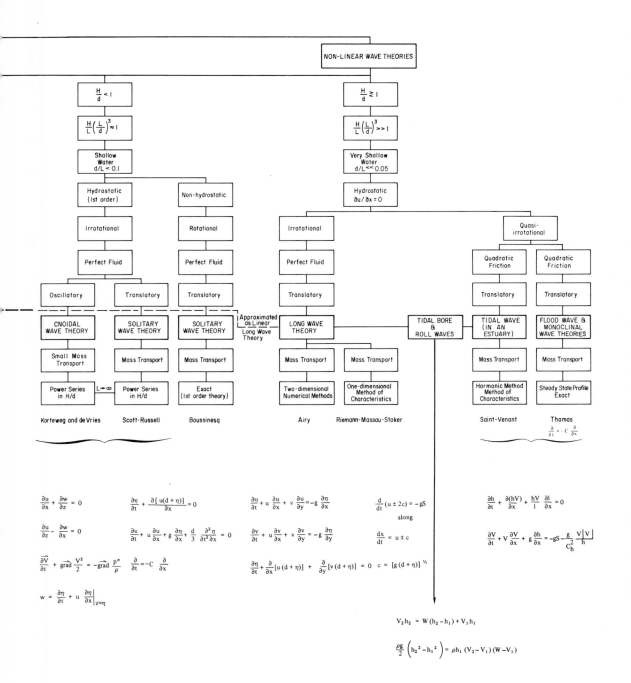

$$\partial^2 \Phi / \partial x^2 + \partial^2 \Phi / \partial y^2 = \nabla_1^2 \Phi = -k^2 \Phi$$

$$(12\text{-}22)$$

and

$$\partial^2 Z / \partial z^2 = +k^2 Z \qquad (12\text{-}23)$$

Therefore $Z(z)$ may be determined as

$$Z(z) = A \sin k \,(kz) + B \cosh (kz) \qquad (12\text{-}24)$$

The boundary conditions require that the normal velocity vanish on the rigid bottom, z=−h,

$$\partial Z / \partial z = 0 \qquad (12\text{-}25)$$

and by Equation 12-14 on the free surface, z=0.

$$-\omega^2 Z + g \,\partial Z / \partial z = 0 \qquad (12\text{-}26)$$

These conditions are homogeneous and will allow for the solution of Z only to within an arbitrary constant while restricting the values of k possible for non-trivial solution. The result is

$$Z = \cosh \,[\,k\,(h + z)\,]\,/\cosh (kh) \qquad (12\text{-}27)$$

and

$$\omega^2 = g \cdot k \cdot \tanh (kh) \qquad (12\text{-}28)$$

where the arbitrary constant has been absorbed into Φ.

Equation 12-22 on Φ is a Helmholtz equation, which also arises in two-dimensional time-harmonic acoustic problems. The boundary condition for Φ on an obstacle will be determined from the corresponding boundary condition on ϕ, provided that this condition is independent of z. For scattering from a rigid cylinder, not necessarily circular, whose axis is parallel to the z axis and which extends from the bottom z = −h through the surface z = 0, this will be the case. The condition on Φ for a rigid scatterer becomes

$$\partial \Phi / \partial n = 0 \qquad (12\text{-}29)$$

on the curve Γ, representing the intersection of the cylinder and the surface z = 0.

While the analogy does not carry over directly for the transient case, a modified version of this integral formulation may be applied there as well, as will be seen shortly.

The corresponding linearized long wave-shallow water problem, however, does go over to an analogous acoustic case for either time harmonic or transient problems as long as the depth h is constant. The governing equation there is

$$\partial^2 \zeta / \partial t^2 = g \cdot h \cdot \nabla_1^2 \zeta$$

which again is the two-dimensional wave equation which governs acoustic problems, reducing to a Helmholtz equation for the time-harmonic case.

Clearly then, for those problems involving constant depth that are described by either linearized transient or time-harmonic long wave-shallow water theory or time-harmonic small amplitude theory, any technique used for solving two-dimensional wave equations, e.g., in acoustics, applies (see Chapter 10).

For regions whose boundaries Γ fall into the category of separable geometries, a useful technique which has been widely applied in acoustic scattering is separation of variables or eigenfunction expansions. Such problems would include scattering of surface waves by circular cylinders, ellipsoidal cylinders, et cetera. This approach is clearly "classical" and depends on computers only in the final determination of numerical values.

As mentioned above, another approach applicable for arbitrary geometries converts the differential equation into an integral equation. This involves only values of the unknown dependent variables on the scattering surface and has been used on two-dimensional time-harmonic scattering problems for both rigid and free boundary condition fixed geometries such as circular cylinders, ellipsoidal cylinders and three-leafed rose cylinders.[8] A corresponding transient solution has been found for a cylinder of square cross section scattering a plane incident step wave. While the corresponding water wave problem (i.e., transient linear shallow water waves scattered by a square "island") would require a continuous incident wave rather than a unit step, the procedure would not be significantly altered. This acoustic problem was solved for rigid, free and impedance boundary conditions.[9,10,11]

While the first boundary condition would be most suitable in actual water wave scattering problems, the third condition in the water wave case is of interest as well in that it requires that the excess pressure (i.e., above hydrostatic) be proportional to the normal velocity at the boundary. This would correspond to a structure which only partially reflects energy while absorbing some as well.[12] This condition may be taken as an approximation to a permeable wall, provided that either no portion of the wave energy is passed through the obstacle or that this transmitted energy is "reintroduced" into the problem as an additional source field on "the other side." While such a question as "transmission" might not arise for a scatterer with a closed cross section, it could be significant in the treatment of a line scatterer, e.g., a semi-infinite permeable breakwater. Although this method is theoretically independent of computer capabilities, in practice it is far more computer-oriented than the method of separation of variables mentioned above. While some solutions to integral equations of this type are obtained by orthogonal series expansions where the coefficients are determined by some form of "least-squares" fit (see Reference 13), the vast majority of applications of this approach have relied on direct numerical solutions which would be literally impossible without a large computing facility.

Since the discussion of the integral equation approach to acoustics problems is available in Chapter 10, there seems little point in repeating it here. It might be mentioned at this point, however, that a similar integral equation approach has been used in Reference 14 to study the response of constant depth harbors of arbitrary shape to periodic plane incident waves. This integral equation approach may also be applied to the transient constant depth, small amplitude theory provided that another Green's function is introduced. This does appear to be worth discussing here, although an equivalent form is presented in Reference 2. This Green's function was derived in Reference 15 and is chosen to satisfy

$$\nabla^2 G(x, y, z, t; x_o, y_o, z_o, t) = \frac{\delta(x - x_o)\,\delta(y - y_o)\,\delta(z - z_o)}{4\pi} \quad (12\text{-}30)$$

in the region of interest where δ is the Dirac delta function and G satisfies

$$\partial^2 G / \partial t^2 + g\, \partial G / \partial z = 0 \quad (12\text{-}31)$$

on $z = 0$

$$\partial G / \partial z = 0 \quad (12\text{-}32)$$

on $z = -h$ with initial conditions

$$G = \partial G / \partial t = 0 \quad (12\text{-}33)$$

at $t = \tau$, i.e., G is the solution to the problem of a point source applied at $t = \tau$ and kept on thereafter. In addition, G and its derivatives will vanish as the independent variables go to ∞, as would be expected in a point source solution.

A scattered potential $\hat{\phi}$ may be introduced as the difference between the total field and the incident field potentials ϕ and ϕ_w respectively. Clearly, $\hat{\phi}$ satisfies

$$\nabla^2 \hat{\phi} = 0 \quad (12\text{-}34)$$

on $z = 0$

$$\partial^2 \hat{\phi} / \partial t^2 + g\, \frac{\partial \hat{\phi}}{\partial z} = 0 \quad (12\text{-}35)$$

on $z = -h$

$$\partial \hat{\phi} / \partial z = 0 \quad (12\text{-}36)$$

and

$$\partial \hat{\phi} / \partial n = -\partial \phi_w / \partial n \quad (12\text{-}37)$$

on the scattering surface. Furthermore, $\hat{\phi}$ will vanish at finite time for points infinitely far away from the scattering surface and will also vanish for time $t \leqslant 0$ (i.e., for t less than or equal to zero, the total field is equal to the prescribed incident field, ϕ_w).

Then, the usual application of Green's theorem, using G and $\partial\hat{\phi}/\partial t$ with an outward normal, n , gives

$$\epsilon\, \partial\hat{\phi}/\partial t = $$

$$\int_S \left[G\, \partial^2\hat{\phi}/\partial t\partial n - \partial\hat{\phi}/\partial t \cdot \partial G/\partial n \right]\, dS \qquad (12\text{-}38)$$

where $\epsilon = 1, 1/2, 0$ depending on whether the field point is within the volume of integration, on the surface S or outside the volume of integration. S includes the upper surface S_F (z=0) the lower surface S_B (z = −h) the rigid scattering surface S_R and S_∞ a "closing" surface at "∞." The integral over S_B vanishes because of the boundary conditions and that over S_∞ because of the behavior of G and ϕ there. Using the case where the field point lies on S_R Equation 12-38 gives

$$\frac{1}{2}\, \frac{\partial\hat{\phi}}{\partial t} = $$

$$\int_{S_F} \left[G\, \frac{\partial^2\hat{\phi}}{\partial t\partial z} - \frac{\partial\hat{\phi}}{\partial t} \cdot \frac{\partial G}{\partial z} \right]\, dS +$$

$$\int_{S_R} \left[G\, \frac{\partial^2\hat{\phi}}{\partial t\partial n} - \frac{\partial\hat{\phi}}{\partial t}\, \frac{\partial G}{\partial n} \right]\, dS \qquad (12\text{-}39)$$

Considering the integral of this equation over time from t = 0 to t = τ gives

$$\frac{1}{2}\, \hat{\phi}(x,y,z,\tau) = $$

$$\int_{S_F} \int_0^\tau \left[G\, \frac{\partial^2\hat{\phi}}{\partial t\partial z} + \frac{1}{g}\, \frac{\partial^2 G}{\partial t^2} \cdot \frac{\partial\hat{\phi}}{\partial t} \right]\, dt dS \; +$$

$$\int_{S_R} \int_0^\tau \left[G\, \frac{\partial^2\hat{\phi}}{\partial t\partial n} - \frac{\partial\hat{\phi}}{\partial t}\, \frac{\partial G}{\partial n} \right]\, dt dS \qquad (12\text{-}40)$$

The first integral may be integrated by parts to give

$$\int_{S_F} \left\{ G\, \frac{\partial\hat{\phi}}{\partial z}\bigg|_0^\tau + \frac{1}{g}\, \frac{\partial G}{\partial t}\, \frac{\partial\hat{\phi}}{\partial t}\bigg|_0^\tau - \int_0^\tau \frac{\partial G}{\partial t} \left[\frac{\partial\hat{\phi}}{\partial z} + \frac{1}{g}\, \frac{\partial^2\hat{\phi}}{\partial t^2} \right]\, dt \right\}\, dS$$

which is zero, since $\partial\hat{\phi}/\partial z$ and $\partial\hat{\phi}/\partial t$ are zero on S_F for t = 0; G and $\partial G/\partial t$ are zero for t = τ; and the boundary condition, Equation 12-35 on $\hat{\phi}$ must be satisfied.

This leaves an inhomogeneous integro-differential equation on ϕ for points only on the scattering surface S_R .

$$1/2\, \hat{\phi}(x,y,z,\tau) = $$

$$\int_{S_R} \int_0^\tau \left[G\, \partial^2\hat{\phi}/\partial t\partial n - \partial\hat{\phi}/\partial t \cdot \partial G/\partial n \right]\, dt dS \qquad (12\text{-}41)$$

While this may not be a particularly appealing integral equation, since G itself involves an (known) integral of Bessel functions, the governing

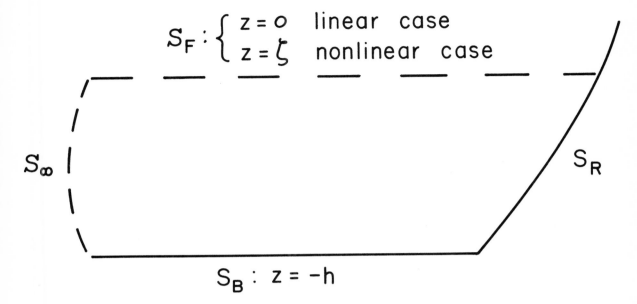

Figure 12-2. Geometry for integral equation approach.

equations have been reduced to an expression using values only on the scattering surface which can represent a significant reduction in the effort required for solution of the problem. Furthermore, this approach can be made to apply to variable depth geometries simply by extending the definition of S_R to include that portion of the bottom surface with variable depth (Figure 12-2). An investigation is presently underway to examine the usefulness of this approach to specific problems by means of approximate or numerical solutions of this integro-differential equation.

A similar treatment may be carried through for the nonlinear "irrotational" theory, which uses the complete nonlinear boundary condition at the free surface. This time, however, a time independent Green's function is chosen as a point source which satisfies boundary conditions on the bottom surface and the scattering surface without any restriction on the free surface. The only other restriction on G is that it and its first derivatives vanish for points infinitely far away from the source point, e.g.,

$$\nabla^2 G = \frac{1}{4\pi} \delta(x - x_o)\, \delta(y - y_o)\, \delta(z - z_o)$$
$$(12\text{-}42)$$

$$\partial G/\partial n = 0 \quad \text{on } z = -h \qquad\qquad (12\text{-}43)$$

$$\partial G/\partial n = 0 \quad \text{on } S_R \quad \text{extended to bound an infinite region above } z = h$$
$$(12\text{-}44)$$

While the Green's function in this case must be determined separately for each scattering geometry, this represents a classical linear problem for a simple equation—Laplace's equation—and simple boundary conditions. For example, two-dimensional solutions could be determined by conformal mapping. Assuming that the incident wave is finite in duration and has no discontinuity in amplitude, the total potential ϕ may be used in this derivation.

Repeating Green's formula, using ϕ and G

$$\int_V \left[\phi \nabla^2 G - G \nabla^2 \phi \right] dV =$$

$$\epsilon \phi =$$

$$\int_S \left[\frac{\partial \phi}{\partial n} \cdot G - \frac{\partial G}{\partial n} \cdot \phi \right] dS \qquad (12\text{-}45)$$

Again, S consists of S_∞, S_F, S_B and S_R (see Figure 12-2); but now the integrals over S_∞, S_B and S_R all vanish, leaving for a field point on S_F

$$\frac{1}{2} \phi(x, y, z \text{ on } S_F, t) =$$

$$\int_{S_F} \left[G \frac{\partial \phi}{\partial n} - \phi \frac{\partial G}{\partial n} \right] dS \qquad (12\text{-}46)$$

where, of course, the location of S_F is unknown, i.e., at z = ζ (x, y, t). Boundary conditions on this surface relate ζ, $\partial\phi/\partial n$ and ϕ; i.e., on z = ζ (x,y,t).

$$\partial\phi/\partial t + \frac{1}{2} (\nabla \phi)^2 + g \zeta = -p_A/\rho \qquad (12\text{-}47)1$$

$$\partial\zeta/\partial t + \nabla \phi \cdot \nabla \zeta - \partial\phi/\partial z = 0 \qquad (12\text{-}48)$$

Therefore, there are three equations on three unknowns: ϕ (x,y,ζ,t), $\partial\phi/\partial n$ (x,y,ζ,t) and ζ(x,y,t). Given an initial configuration, i.e., ϕ and ζ at t = 0, a time-marching numerical solution could be obtained using Equation 12-47 to determine $\partial\phi/\partial t$ and, therefore, ϕ at the next step–t + $\triangle$ t; Equation 12-48 to determine $\partial\zeta/\partial t$ and, therefore,ζ ; and, finally, Equation 12-46 to determine $\partial\phi/\partial n$ once ϕ and ζ are known at t + $\triangle$ t.

These two integral equation approaches have been presented for consideration as possible solution methods to problems involving complicated geometries, although examples would be required to check the usefulness of these formulations. The question of uniqueness of solution, for example, arises in the second method, where the surface S_R —once it is beyond the volume containing the fluid—may be defined in a non-unique fashion, implying a non-unique Green's function. It must be established, then, that the solution is still unique.

An extremely useful group of problems in any area in which numerical and/or approximate solutions are common are those for which exact solutions can be found. These may then serve as test cases for numerical and/or approximate schemes or even "building blocks" for the (approximate) solution of more complicated problems (e.g., the solution for a finite breakwater diffraction problem may be approximated by the superposition of the solutions for two semi-infinite breakwaters).

The vast majority of problems solved exactly are those involving planar obstacles, such as a semi-infinite breakwater,[16] a submerged barrier,[17] a finite depth surface barrier,[18] etc., scattering time harmonic small amplitude waves.

The solution of the semi-infinite breakwater problem is particularly interesting, since it can be obtained in closed form.

The original solution for normal and oblique angles of incident wave motion relative to the vertical semi-infinite breakwater, e.g., Figure 12-3, appears to have first been given in Reference 16 by recognizing the analogy to the well-known solution to the half plane diffraction problem due to Sommerfeld. The same solution has been repeated by several authors who have attempted to recast the solution in more acceptable engineering forms, e.g., References 19, 20 and Figure 12-4. Since this problem is a very good representation of an actual breakwater, the solution has proved useful in practice. It is limited, however, by the assumption of constant depth which cannot allow for the possibility of wave refraction, which may be a significant feature in many problems. This solution has also proved useful in the investigation—at least, ap-

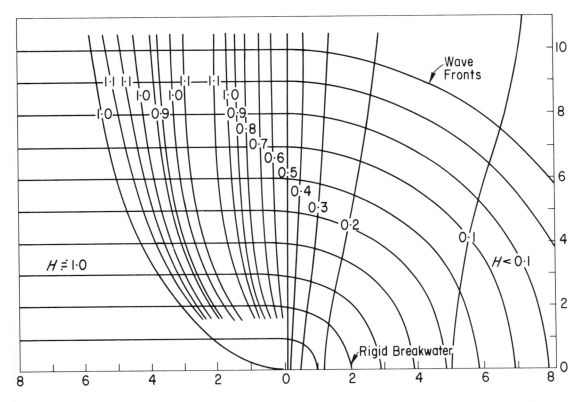

Figure 12-3. Wave fronts and contour lines of maximum wave heights in the lee of a rigid breakwater, the waves being incident normally (after Penney and Price[16]).

proximately—of related problems, such as diffraction of time-harmonic incident waves by an infinite breakwater with a gap[16,21] or a finite length breakwater[16,20] if the ends of the breakwater are at least several wave lengths apart. Sommerfeld extended his method to treat the problem of scattering of waves by a wedge of angle $2\pi m/n$, where m and n are integers, and such solutions have also been found useful in water wave problems.[22]

The other problems mentioned above seem to fall into the general category of Wiener-Hopf problems, and that method has been particularly useful for asymptotic (i.e., large distance from the scatterer) solutions. It does not seem appropriate here to go into the details of this method, and the read-er is referred to such standard texts as Reference 23.

The most common group of computer oriented solution techniques are those based on direct finite difference approximation of the governing equations of motion. Reference 24 contains several sections which discuss these approaches with particular emphasis on the questions of using either Eulerian or Lagrangian formulations (or even some combination of the two), i.e., equations written in terms of a reference which is fixed in the laboratory system or equations written in terms of a reference which is fixed to the fluid, respectively.

The Eulerian system has the advantage of having a fixed mesh for finite differencing, but has

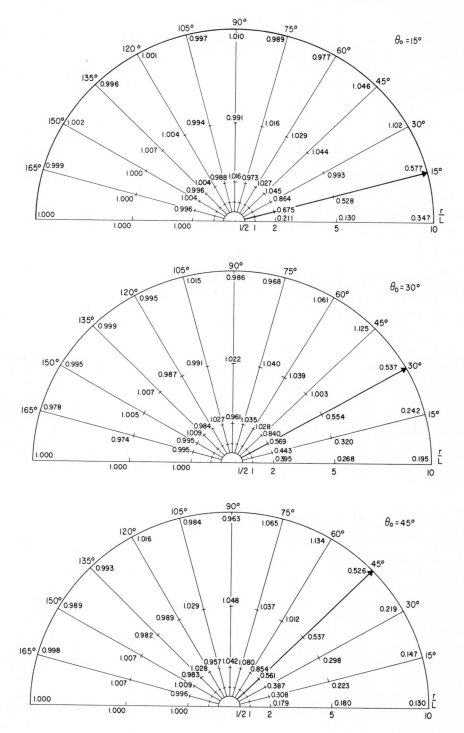

Figure 12-4.   Diffraction coefficients as a function of incident (after Wiegel[19]).

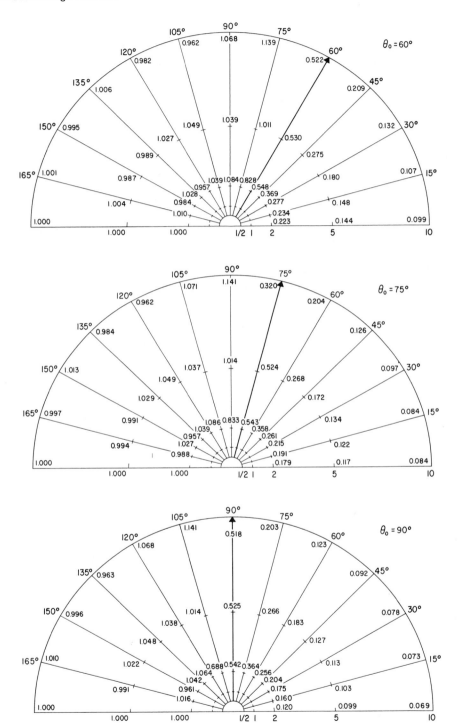

Figure 12-4. Wave angle for semi-infinite rigid impermeable breakwater (after Wiegel[19]).

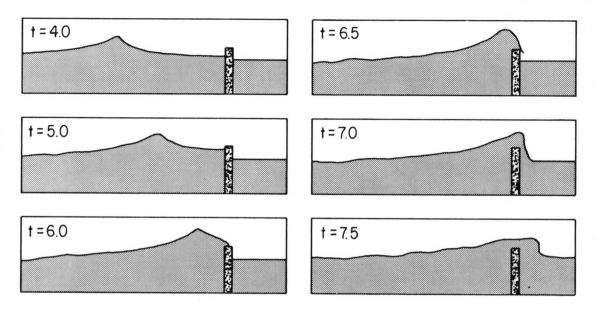

Figure 12-5. Wave on a breakwater (after Welch, Harlow, Shannon and Daly[27]).

difficulty in representing material interfaces (e.g., in our problems the free surface) which move with the fluid. The Lagrangian system, on the other hand, has difficulty in handling severe distortions of the original mesh. For the linearized problems which form a large part of water wave investigations, these systems are essentially the same.

Other questions of importance concerning finite difference techniques, such as stability and convergence, are also discussed in Reference 24 with specific application to hydrodynamics, in Reference 25 and in standard texts on numerical analysis, such as Reference 26.

A final group of techniques applicable to water wave problems, which are even more computer oriented than those mentioned so far, are the particle-in-cell methods (and their "descendents"). In particular, for free surface problems, the MAC (marker and cell) methods[27] has been used to solve transient water wave problems. The equations used are the full nonlinear equations, which allow for treatment of such interesting problems as breaking of waves. Viscous effects are included, and irrotationality is not required.

This approach uses a fixed mesh (Eulerian) coordinate system with massless "markers," which move with the fluid and which are used to follow the free surface. The markers may be used to distinguish the free surface from the interior of the fluid, thereby indicating the location at which the free surface boundary conditions are to be applied. The method developed in a "trial-and-error" sense, however, and some of the steps which appear arbitrary are really quite vital to the successful application of this approach. One such step is the recasting of the governing differential equations into a "conservation" form, whereby the corresponding finite difference approximations also "conserve" the required properties. For example, conservation of mass

$$\nabla \cdot \overline{u} = D = 0 \qquad (12\text{-}49)$$

is already in the appropriate form, while conservation of linear momentum

$$\frac{\partial \overline{u}}{\partial t} = -(\overline{u} \cdot \nabla)\overline{u} - \nabla p/\rho + \nu \nabla^2 u + \overline{G}$$

$$(12\text{-}50)$$

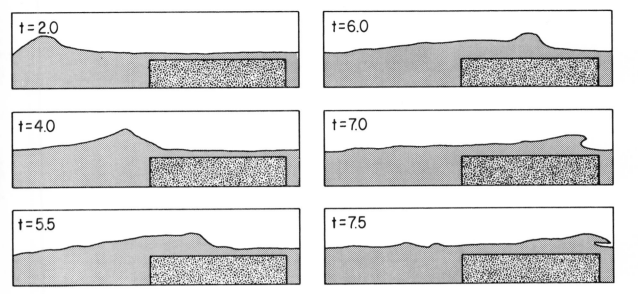

Figure 12-6. Wave on a reef (after Welch, Harlow, Shannon and Daly[27]).

where $\overline{G} = \hat{g}\hat{k}$ = gravity body force vector is combined with the conservation of mass equation to give

$$\frac{\partial \overline{u}}{\partial t} = - \nabla \cdot (\overline{u}\,\overline{u}) - \nabla p/\rho + \nu \nabla^2 \overline{u} + \overline{G}$$

$$(12\text{-}51)$$

to put the right-hand side in the form of a divergence plus a conservative body force, i.e.,

$$\frac{d}{dt} \int_V \overline{u}dV = \int_S \left\{ - (\hat{n} \cdot \overline{u})\,\overline{u} - \hat{n}\,p/\rho + \nu\,(\hat{n} \cdot \nabla)\,\overline{u} \right\} dS + \int_V \overline{G}\,dV$$

$$(12\text{-}52)$$

such that the only contribution to momentum (proportional to $\overline{u}$ for constant density) in a volume V comes from surface fluxes plus a body force. The original momentum equation could not be represented in this form. While both forms are

correct as differential equations, the latter form is superior when a finite difference approximation is used. Another such step involves the location of the fixed mesh points at which the dependent variables are to be specified. These are taken at the midpoint of the faces joining neighboring cells rather than at the center of an individual cell. This allows the conservation of mass equation, for example, to be written for a given cell without involving cells other than its immediate neighbors. If this were not done, the "far distant" contributions would complicate the calculations and might even introduce instabilities.

The computational scheme is a time-stepping, or "movie frame," technique in that calculations are performed over one finite but small time step starting from certain initial conditions. After all variables have been calculated at the new time and markers moved to indicate the new position of the free surface, the final state is then used as an initial condition for the next time step. The dependent variables at the fixed mesh points are found by solving the finite difference form of Equations 12-49, 12-51 with free surface boundary conditions applied at those cells indicated as being at the free surface by the markers. (See Reference 28 for a film version of these results.) The specific form of the free surface boundary conditions for highly viscous flows (low Reynolds number) has caused some difficulty;[29] but, for most water wave problems, this would not be a significant effect. Results of two calculations by this method are shown in Figure 12-5 and 12-6 for a solitary wave on a breakwater and on a reef, respectively.

In conclusion, some general approaches to the solution of gravity wave diffraction problems have been presented. These methods are based to a large part on the use of digital computers e.g., direct finite difference methods, integral equation methods and marker-in-cell methods and may be somewhat less familiar than many of the "classical methods," e.g., separation of variables and eigenfunction expansions. Two points bear mention, however. First, the real ocean is not anywhere as deterministic as assumed in this discussion; and, while frequently a solution to a deterministic problem will aid in the solution of the more realistic stochastic problem, extensions to stochastic problems in regard to both incident wave form and topography are required. Second, many realistic physical effects—such as material viscosity, bottom friction, wind surface stresses, variable atmospheric pressure and surface tension—have been disregarded. While many of these may be included in some of the methods described above, further research appears necessary on their significance. One of the primary difficulties, of course, in deciding which effects are and are not significant in the real ocean is that in situ studies are difficult, while model studies may not always represent the true physical picture. Although laboratory model studies continue to play a large role in the development of water wave theory, as computing capabilities continue to increase and costs decrease, it nevertheless appears that computer simulated studies such as those of Reference 25 will certainly become a more common and possibly even become one of the most significant tools in water wave research.

List of Symbols

G	Green's function
g	Acceleration due to gravity
h	Water depth
k	Wave number
k	Unit vector in z direction
n	Normal direction
p	Pressure
S	Surface of integration
S_F	Upper (free) surface
S_R	Rigid scattering surface
S_B	Bottom surface
S_∞	Closing surface at infinity
t	Time
$\overline{u}$	Velocity field
(u,v,w)	Components of velocity
x,y	Horizontal cartesian coordinates
z	Vertical cartesian coordinate
Z	Function of z in Equation 12-21
ϵ	Constant used in Equation 12-38
Γ	Boundary curve
ϕ	Velocity potential
ϕ	Scattered velocity potential
ϕ_w	Incident velocity potential
Φ	Function in Equation 12-21
τ	Time variable
$\overline{\overline{\tau}}$	Stress tensor
ρ	Fluid density
ω	Frequency
ζ	Disturbed surface level

References

1. Lamb, H. *Hydrodynamics.* 6th edition, Dover Publications, N.Y.C., 1945.45.

2. Stoker, J. J. *Water Waves.* Interscience Publishers, N. Y. C., 1957.

3. Wehausen, J. V. and E. V. Laitone, "Surface Waves," *Handbuch der Physik.* S. Flugge (ed.), Vol. IX, Fluid Dynamics III, Springer-Verlag, Berlin, 1960.

4. John, F. "On the Motion of Floating Bodies," *Comm. Pure. App. Math,* 2, No. 1 (1949), 13-57.

5. Eckart, C. *Hydrodynamics of the Ocean and Atmospheres.* Pergamon Press, N. Y. C., 1960.

6. Kinsman, B. *Wind Waves.* Prentice Hall, Inc., Englewood Cliffs, N. J., 1965.

7. Le Mehaute, B., D. Divoky and A. Lin. "Shallow Water Waves: A Comparison of Theories and Experiments," Tetra Tech, Inc., Pasadena, California.

8. Banaugh, R. P. and W. Goldsmith. "Diffraction of Steady Acoustic Waves by Surfaces of Arbitrary Shape," *Jour. Acous. Soc. Am.,* 35, No. 10 (October, 1963), 1590-1601.

9. Friedman, M. B. and R. P. Shaw. "Diffraction of a Plane Shock Wave by an Arbitrary Rigid Cylindrical Obstacle," *Jour. App. Mech.,* 29, No. 1 (March, 1962), 40-46.

10. Shaw, R. P. and M. B. Friedman, "Diffraction of a Plane Shock Wave by a Free Cylindrical Obstacle at a Free Surface," *4th U. S. National Congress of Applied Mech.* (June, 1962), 371-379.

11. Shaw, R. P. "Diffraction of Pulses by Obstacles of Arbitrary Shape with an Impedance Boundary Condition," *Jour. Acous. Soc. Am.,* 44, No. 4 (October, 1968), 1062-1068.

12. Le Mehaute, B. "The Perviousness of Rock Fill Breakwaters to Periodic Gravity Waves," (in French), *La Houille Blanche,* 12, No. 6 (December, 1957), 903-919.

13. Sharma, D. L. "Scattering of Steady Elastic Waves by Surfaces of Arbitrary Shape," *Bull. Seis. Soc. Am.,* 57, No. 4 (August, 1967), 795-812.

14. Hwang, L. S. and B. Le Mehaute. "On the Oscillations of Harbors of Arbitrary Shape," Tetra Tech Report NVO-289, Tetra Tech Inc., Pasadena, California, October 1968.

15. Finkelstein, A. B. "The Initial Value Problem for Transient Water Waves," *Comm. on Pure and App. Math,* 10, No. 4 (1957), 511-522.

16. Penney, W. G. and A. T. Price. "The Diffraction Theory of Sea Waves and the Shelter Afforded by Breakwaters," Part I of "Some Gravity Wave Problems in the Motion of Perfect Liquids," *Phil. Trans. Roy. Soc.,* Part A, 244 (1952), 236-253.

17. Faulkner, T. R. "The Diffraction of an Obliquely Incident Surface Wave by a Submerged Plane Barrier," *Z. A. M. P.,* 17 (1966), 699-707.

18. _____."The Diffraction of an Obliquely Incident Surface Wave by a Vertical Barrier of Finite Depth," *Proc. Camb. Philo. Soc.,* 62 (1966), 829-838.

19. Weigel, R.L. "Diffraction of Waves by Semi-Infinite Breakwater," *ASCE,* 88, No. HY1 (January 1962), 27-44'

20. Putnam, J. A. and R. S. Arthur. "Diffraction of Water Waves by Breakwaters," *Trans. Am. Geophy. Un.,* 29, No. 4 (August 1948), 481-490.

21. Blue, F. L. and J. W. Johnson. "Diffraction of Water Waves Passing through a Breakwater Gap," *Trans. Am. Geophy. Un.,* 30, No. 5 (October 1949), 705-718.

22. Sebekin, B. I. "Diffraction of Surface Waves on a Wedge," *Izu. Acad. of Sciences, USSR, Mechanics of Fluids and Gases,* 5 (1966).

23. Carrier, G. F., M. Krook and C. E. Pearson. *Functions of a Complex Variable.* McGraw Hill Book Co., N. Y. C., 1966.

24. Alder, B., S. Fernbach and M. Rotenberg (ed.). *Methods in Computational Physics, Vol. 3, Fundamental in Hydrodynamics.* Academic Press, N. Y., 1964.

25. _____ *Methods in Computational Physics, Vol. 4, Applications in Hydrodynamics.* Academic Press, N. Y., 1965.

26. Fox, L.,*Numerical Solution of Ordinary and Partial Differential Equations.* Addison-Wesley Publishing Co. Inc., Reading, Mass., 1962.

27. Welch, I. E., F. H. Harlow, J. P. Shannon and B. J. Daly. "The MAC Method," LA 3425 Revised, Univ. of Cal., Los Alamos Scientific Laboratory, January 1969.

28. *Numerical Calculations in Hydrodynamics,* Film Y–204, Report Library, Los Alamos Scientific Laboratory, Univ. of Cal., Los Alamos, New Mexico, 1969.

29. Hirt, C. W., and J. P. Shannon. "Free Surface Stress Conditions for Incompressible Flow Calculations," *Jour. Comp. Phy.,*2 (1968), 403-411.

30. LeMéhauté, B. *An Introduction to Hydrodynamics and Water Waves, Volume II: Water Wave Theories.* ESSA Tech. Rep. ERL 118-POL 3-2, *U. S. Govt. Printing Office,* July 1969.

part 8: willard j. pierson, jr.

Lagrangian Studies of Waves

The Radscat Concept

Developments in Numerical Wave Forecasting

Spacecraft Altimetry and Ocean Geoid

Author of approximately 25 publications—books, journal articles, papers—Prof. Willard J. Pierson, Jr., has his B.S. (1944) from the University of Chicago and his Ph.D. (1949) from New York University.

He has been a faculty member in the Department of Meteorology and Oceanography at NYU since 1949, beginning as an instructor, becoming an associate professor in 1959 and a professor in 1962. He teaches undergraduate and graduate courses. His areas of specialization are geophysical random processes, ocean waves, air sea boundary processes and turbulence.

Prof. Pierson belongs to the American Meterological Society, American Geophysical Union (president, Section on Oceanography, 1969-1971), American Association for the Advancement of Science, Society of Naval Architects and Marine Engineers (Seakeeping Panel H-7), Sigma Xi (National Lecturer, fall 1967) and the Marine Technology Society.

Also, he is a member of the Earth Resources Advisory Subcommittee, NASA, and is on the board of directors (and senior scientist) of Oceanics, Inc., Plainview, N.Y. He has been consultant to numerous companies.

From 1944-46 Prof. Pierson served as U.S. Air Force Weather officer.

13

lagrangian studies of waves

Introduction

The material for this review is well described in the paper by Chang[1]. For this review, it will suffice to reemphasize some of the more striking results of Chang and to describe some of the results obtained in this subject since the work of Chang.

Properties of the Solution

The full perturbation solution to second order for irrotational particle motions in long crested random waves both at the surface and at all depths in deep water was found by Chang. This means that if the solution that she obtained is substituted into the original complete nonlinear Lagrangian equations, all terms will vanish identically except those of third, or higher order.

An interesting property of the second order terms in the solution is that they involve only difference frequencies. Terms equivalent to the second harmonic terms that arise in analyses of the Eulerian equations are absent. The difference frequency effects turn out to be very large and

extend to great depths. They can be explained intuitively as the effect of large mass transports where the wave groups are high and weaker mass transports where the wave groups are low. This produces a convergence-divergence field with variations over tens of surface wavelengths (depending on the spectrum) at and near the surface and areas of vertical motion with return flows that extend to great depths.

Practical Applications

The practical applications of the work of Chang have been investigated by Chin and Pierson[2] as reported at the Sixth Military Oceanography Symposium held in Seattle in the spring of 1969. The computation of the surface drift was extended to short crested waves. A computer program was written that obtained the surface drift given the hindcasted (or forecasted) wave spectra obtained by methods described by Marks et al.[3]

A typical flow field for a certain map time as produced by all wave components with frequencies less than $\omega = 1.03$ radians per second was found. This high frequency cutoff makes the results representative of an average over a depth of about 3 meters. For a wind field with a small area of 40 knot winds and other areas of 30 knot winds, the Stokes' drift was as high as 15 cm/sec (3/10 of a knot) and compared favorably with the kinds of numbers one gets from computing Ekman transports. The results also compared well with many surface current measurements as compared with wind speed. Considerable work is still needed in determining the relative importance of Stokes' transport and Ekman drift.

Another aspect of this report concerned the motions of what would be levels of constant depth in a fluid at rest. The 30 meter level can move up and down 4 meters and the 70 meter level, 1 meter, due to these Lagrangian effects alone. These vertical motions can have the time and space scales of internal waves and care is needed to distinguish such effects from internal waves.

References

1. Chang, M. S.: "Mass Transport in Deep-water Long-crested Random Gravity Waves," *J. Geophys. Res.,* 74, No. 6 (1969).

2. Chin, H. and W. J. Pierson. "Mass Transport and Thermal Undulations Caused by Large and Small Scale Variability in the Stokes' Mass Transport Due to Random Waves." *Proceedings of the 6th U.S. Navy Symposium on Military Oceanography, Seattle, Washington.* 1969.

3. Marks, W., T.R. Goodman, W.J. Pierson, Jr., L. J. Tick and L. A. Vassilopoulos. "An Automated System for Optimum Ship Routing," *Trans. Soc. Naval Arch. and Marine Engineers.* 76 (1968), 22-25.

14

the radscat concept

Moore and Pierson[1] have proposed a composite microwave-radiometer-scatterometer on a NIMBUS spacecraft for the study of winds and waves, precipitation, sea ice and ocean surface temperature. The summary of this proposal is quoted below.

Introduction and Summary

The mission of the proposed experiment is to establish the feasibility of an operational system that will upgrade meteorological and oceanographic prediction systems by providing information on the waves on the ocean, the winds over the ocean, the precipitation over the ocean, the movement of sea ice, and perhaps the ocean surface temperatures. The experiment is proposed to be flown on the Nimbus F Spaceflight.

We believe that, after a calibration and testing period of several months, the data output from this experiment can be directly incorporated into the numerical prediction methods of ESSA and the U.S. Navy on a routine basis.

Observations over the past year in the NASA Earth Resources Program, coupled with further analysis of previous measurements lead to the conclusion that centimeter-wavelength radar scattering at angles between 20°

and 50° from the vertical is directly related to wind and wave conditions for fully developed seas, and that, for partially developed seas, the radar return is related to the high-frequency part of the wave spectrum which is directly determined by the local wind. The theory of radar sea return suggests that a relation exists between smaller structures on the ocean surface and the magnitude of the radar return. Since smaller waves are due to the local wind field, a direct relation between radar return and local wind field appears possible. Recent theoretical work suggests that a reasonably accurate theory for radar sea return may soon become available and explain measurements made by the NASA surface truth missions.

The theory of microwave emission from the surface of the earth is not nearly as well developed as radar theory, although the theory of emission from precipitation is well known. Radiometric measurements over the sea under carefully controlled conditions are even less developed. We believe, nevertheless, that the proposed composite instrument will provide significant information about the earth's surface not available from radar alone, and, in the absence of adequate knowledge of the surface measurements capability of the radiometer, its precipitation measurement capability justifies its inclusion in the proposed instrument.

Emissivity and radar differential scattering cross section are controlled by the same factors: surface roughness and dielectric properties. The exact form of the relationship between them has not been established for the sea. There is a strong possibility, however, that measurements of the radar scattering and thermal emission together may permit separation of temperature and emissivity effects so that the radiometer can be used as a temperature measuring instrument for the ocean. Unlike infrared devices, it will respond to temperature to a depth of several millimeters.

The advance and retreat of the Arctic and Antarctic pack ice can be mapped with the proposed instrument with applications to shipping, meteorology, and wave forecasting. Radar has clearly shown its capability to detect and differentiate sea ice, and radiometry has been used also for detecting ice. The combination should permit better differentiation of ice types. Since the resolution cell is large, the application is to general mapping, not to location of specific ship channels and icebergs. Further, the proposed sensor will make possible mapping of seasonal changes in snow cover and freezing-melting boundaries and various features of the land related to vegetation and ground moisture. Because of the large resolution cell, the true value of this function can only be evaluated with the spaceborne instrument.

The radiometer will permit the calibration of the influence of rain attenuation on the radar, at the same time as the presence of rain will be determined for direct meteorological application. Only in very extensive regions of heavy precipitation will this rain attenuation affect the accuracy of the radar measurements. The possibility of a slight loss of information in such areas is largely compensated by the advantages that are derived from the fact that the frequency of the composite instrument allows the use of a reasonably sized antenna.

The proposed experiment is based on simultaneous measurement of the differential backscattering cross section and the microwave emissivity of the sea surface, on a global scale, with a composite microwave RADiometer-SCATterometer (RADSCAT).

Surface cells of adequate dimensions are explored with a radar technique: short CW pulses, at 10.0 GHz* are sent to an illuminated surface cell. The backscattered energy is captured by a low nose narrowband receiver. The magnitude of the integrated signal is a measure of the differential back-scattering cross section.

A narrow electronically scanned pencil beam illuminates the selected cells in time sequence, moving in steps from left to right, dwelling on each cell for a predetermined length of time. An interlaced surface trace is obtained where the rows of cells are contiguous in the in-track direction and separated by approximately a cell length in the cross-track direction. Each cell is viewed once at a predetermined angle, thus yielding a measure of the differential back-scattering cross section.

The proposed instrument makes use of techniques and technologies that are well within the state of the art. Thus, antennas similar to the proposed electronically scanned phased array antenna are not new. The weight of the antenna is compatible with the Nimbus E Spacecraft and its size permits stowing and deployment in a manner similar to the one intended for Nimbus E. The electronics is all solid-state except for a TWT tube. This tube will be operated at a peak power of less than 20 watts with a duty cycle of approximately 15 percent, and is presently available from at least one TWT manufacturer in space-qualified form. The total average dc power requirement for this experiment is less than 30 watts and is

---

*The exact frequency of the instrument is 9.9 GHz although it is often referred to as 10.0 GHz.

within the power levels allocated to other passive microwave experiments.

Most electronic components, and particularly the front end of the receiver and the transmitter, are mounted on the back of the phased-array antenna, thus keeping transmission lines short and alleviating the thermal burden on the spacecraft sensor ring. The IF portion and the output portion of the receiver, the control and beam steering networks are located in one small portion of a Nimbus F bay.

The following sections of this proposal deal in detail with the results that are expected from this experiment and the manner in which these results will be used for the study of winds and waves over the ocean, sea ice, precipitation and weather prediction. A brief discussion of the instrument is included, summarizing the preliminary design and the engineering feasibility of the proposed composite sensor.

The data on radar scattering cross section as a function of incidence angle and wind-wave conditions that were available at the time of the proposal indicated a strong dependence on wind and waves. These data have been processed by analogue methods. More recent data processed by digitization methods led to the discovery of sources of error in the instrumentation which indicate the need for additional proof of concept. Moreover, independent measurements by NRL over a wide variety of sea conditions do not show as great a dependence on wind and waves at 8 gigahertz as the scatterometer appears to have obtained at 13 gigahertz.

The present status of the theory and observation of scattering from rough surfaces has been given by Moore and Pierson.[2] Both NRL (Guinard and Daley[3]) and NASA (Moore[2]) data show a dependence of radar sea return on wind and wave conditions. However, Guinard and Daley find at 8 gigahertz a much weaker dependence relatively independent of incidence angle, and Moore finds a

variation in the shape of the curves such that a value at 35° differs by 15 db for winds of 12.5 and 49 knots compared to their differences at 10°. These differences could be due to either differences in instrumentation or to actual differences in sea surface roughness at the two frequencies. The radar wavelength at 13 gigahertz is 2.25 cm and water waves at these wavelengths have strange properties as shown by Pierson and Fife.[4] They could be quite steep, for example.

A prototype combination radiometer scatterometer for aircraft use at 10 gigahertz has been built by General Electric. This instrument will for the first time make it possible to obtain radiometric and microwave data from the same resolution cell on the sea surface.

A multiple purpose radar scatterometer, altimeter and passive microwave receiver at 1.3 GHZ will be flown on SKYLAB I in 1972 according to present contracts and plans. The concepts given above can be adequately tested by this instrument.

The writer confidently expects that proof of concept of the combination radiometer-scatterometer will become available and the instrument will eventually be used on a spacecraft as described above.

References

1. Moore, R.K. and W. J. Pierson. "RADSCAT, A Composite Microwave Radiometer-Scatterometer," Proposal submitted to NASA on October 31, 1968.

2. Moore, R. K. and W. J. Pierson, "Worldwide Oceanic Wind and Wave Predictions Using a Satellite Radar Radiometer". AIAA Paper No. 70-310, 1970.

3. Guinard, N. W. and J. C. Daley. "An Experimental Study of a Sea Clutter Model," Submitted to *Proc. IEEE*, 1969.

4. Pierson, W. J. and P. Fife. "Some Nonlinear Properties of Long Crested Waves with Lengths Near 2.44 Centimeters," *J. Geophys. Res.*, 66 (1961), 163-179.

15

developments in numerical wave forecasting

Introduction

Efforts at New York University to develop a computer based numerical procedure for forecasting and hindcasting wind generated waves and swell have been described by Pierson, Tick and Baer.[1] An earlier model for the North Atlantic was completed in 1964 and revised so as to include the spectral wave growth theory of Miles[2] and Phillips[3] according to procedures described by Inoue.[4] This model has been described by Marks et al.[5] Tests of this North Atlantic model have been made and are described in the unpublished work of Bunting and Moskowitz.[6] Some results, nearing completion, that will define the wind fields more accurately have been obtained by Cardone.

In this review, the properties of the North Atlantic model will be described along with results of its evaluation. Then the properties of a global model will be summarized, and, finally, new results by Cardone on the planetary boundary layer will be given.

North Atlantic Model

The North Atlantic model was for a flat earth and curvature effects on propagation were neg-

lected as not being very serious. The planetary boundary layer was assumed to be neutrally stable so that a simple logarithmic wind law could be used. The spectral form for fully developed seas proposed by Pierson and Moskowitz[7] was used, and the growth rate at each spectral band was based on the work of Inoue,[4] who studied the theories of Miles[2] and Phillips.[3]

The wave spectra were defined at over 500 points by 180 numbers representing fifteen frequency bands and twelve direction bands, and each of the 180 fields was propogated in the appropriate direction at the correct group velocity. Wave dissipation for waves traveling against the wind is brought about by an effective eddy viscosity based on the strength of a wind generated sea. A description of this model has been given by Marks et al.[5]

Evaluation by Bunting and Moskowitz

Wave specification tests and simulated wave forecasts were carried out for several versions of the North Atlantic model. Wave specification, or wave hindcasting, consists of computing what the waves would have been like given analyzed

6-hourly pressure fields and ship reports, and wave forecasting consists of using the wave specification results up to a certain map time followed by stored data for the 6, 12, and so on, hourly forecasts of surface pressure fields from which the winds and then the waves are computed.

An evaluation of a selected set of forecasts was made by Bunting and Moskowitz, and the publication of the results is in progress.

Simulated forecasts from 6 hours to 36 hours were made at weather ships A, I, J, K and at Argus Island. There were 29 six-hour forecasts, 41 twelve-hour forecasts, 26 eighteen-hour forecasts, 40 twenty-four-hour forecasts, 29 thirty-hour forecasts, and 25 thirty-six-hour forecasts for a total of 190 forecasts. Some were made using the U.S. Weather Bureau prognostic pressure fields and others were made using the Fleet Numerical Weather Facility prognostic pressure fields, and the waves, as computed from waves hindcasted up to a given map time and forecasted past that map time, were obtained. The forecasts were verified by means of the wave records obtained by either the Tucker shipborne wave recorder, after proper calibration in the spectral domain, or the wave recorder at Argus Island. The report by Bunting and Moskowitz draws many interesting conclusions. The following two tables illustrate some of their results.

Table 15-1
RMS Errors in Wind and Wave Forecasts Using U.S. Weather Bureau Surface Prognostic Charts

| Ship | No. of Obs. | RMS ERROR | | |
| | | Wind | | Waves |
		Degrees	Knots	Sig. Hgt (ft)
A	12	65.1	16.8	11.3
I	15	63.4	9.3	4.7
J	29	70.5	12.3	7.7
K	30	95	7.3	3.3
All Ships	86	77	11.0	6.7
Argus Is.	38	66	5.4	2.7

Table 15-2
RMS Errors in Wind and Wave Forecasts Using Fleet Numerical Weather Facility Surface Prognostic Charts

| Ship | No. of Obs. | RMS ERROR | | |
| | | Wind | | Waves |
		Degrees	Knots	Sig. Hgt (ft)
A	81	66	17	10
J	24	66	12	5
K	17	100	9	3
All Ships	49	79	12	6
Argus Is.	17	74	5	3

The wave height forecasting error is well correlated with errors in the wind velocity forecasts. The particularly poor forecasts at Weather Ship A are probably due to poor wind analyses and forecasts near Greenland, and particular attention to this area can probably reduce this source of difficulty. The actual waves varied in significant height from 15 to 35 feet at the weather ships. The spectra, as a function of frequency, verified well whenever the forecasted and observed significant heights were close.

Global Model

The global model under development will have numerous features that are superior to the North Atlantic model. It will define the wave spectra by 360 numbers instead of 180, and the waves will be propagated on great circle paths.

The greatest improvement will, however, come from improved methods for analyzing and forecasting the wind field in the planetary boundary layer. There will be four times the number of points analyzed by halving the scale of the JNWP grid.

Appendix A describes this model of the planetary boundary layer developed by Cardone, and Appendix B describes the computer programs he has developed to carry out these computations. A report on this aspect of the work has appeared.[8]

Appendix A

Computer-Based Model for the Analysis of the Planetary Boundary Layer by Means of Conventional Data by Vincent Cardone

A computer analysis model has been developed that provides the detailed meteorological input required by state of the art wave specification models. The model utilizes a routinely available ships' weather observations to diagnose the surface and planetary boundary layer wind distribution over the oceans each six hours on a grid system corresponding to the JNWP grid but with the grid spacing halved so as to quadruple the number of grid points.

The analysis procedure starts with the application of a planetary boundary layer model to conventional analyses of sea-level pressure, air temperature and air-sea temperature difference to provide an initial evaluation of the surface boundary layer wind distribution. Developmental research on this aspect of the model has yielded the following results.

1. Though considerable progress has been made in the past decade in our understanding of the distribution of wind in the planetary boundary layer, none of the theoretical models proposed prior to this study could be applied to the non-neutral baroclinic boundary layer over the sea.

2. One of the most successful models was successfully extended to include the effects of a non-neutral stratification and an internally prescribed description of the lower boundary. Baroclinicity was included in a way such that its effects could be specified from conventional analyses of surface air temperature, and the stability influence was formulated so that its effects could be specified from the air-sea temperature difference.

3. The characteristics of the surface boundary layer wind distribution were found to depend significantly upon the geostrophic wind speed, the air-sea temperature difference, the magnitude and orientation of the horizontal temperature gradient and latitude. The ratio of the typical anemometer level wind to the geostrophic wind can range from 40% to 90% over typical ranges of wind speed and air-sea temperature difference. Baroclinicity and air-sea temperature were found to affect the directional characteristics of the boundary layer wind distribution.

4. Existing statistical studies of the speed and directional characteristics of the anemometer level wind over the sea were found to be special cases (in terms of latitude, height at which wind speed was measured, range of wind speeds and stabilities involved and considerations of the average correlation between stability and baroclinicity) of the model. When the model parameters were suitably restricted, the theoretical relationships compared very well with the statistically derived relations.

5. The boundary layer model approach makes possible the calculation of the input parameters required by wave specification models from routinely available prognostic fields.

6. Should spacecraft sources of surface wind information become available, the model may be utilized in reverse to specify the distribution of wind and pressure to the top of the planetary boundary layer from surface layer information alone.

The planetary boundary layer model is consistent, as the surface is approached, with a surface boundary layer model constructed around the similarity profile forms that have been shown to describe the distribution of wind and temperature in the lowest 10-100 meters of atmosphere over land and the sea. The following conclusions could be drawn from research on this aspect of the model.

1. A combination of inferences from recent theoretical work on wave generation and evidence from recent field experiments in wave generation and in turbulence studies over water leads naturally to the conclusion that at least for situations of active wave generation the profile method may be applied to the calculation of the turbulent fluxes in the surface boundary layer over the sea.

2. Within this restriction, the main difference between the surface boundary layers over land and that over the ocean lies in the complexity of the roughness parameter specification for the latter. In this study, the roughness parameter is prescribed internally in terms of physical constants and the surface shearing stress. The proposed relation interpolates between aerodynamically smooth and rough flows and is consistent with a large body of observational evidence.

3. The surface boundary layer model effectively separates the effects of wind speed, stability and anemometer height on the ratio of the surface shearing stress to the anemometer level wind speed. The predictions of the model compare favorably with recently obtained direct measurements of the turbulent Reynolds stress over the open ocean over a wide range of wind speeds and stabilities.

4. A technique has been developed whereby the surface boundary layer wind distribution may be specified from single layer measurements of wind and air temperature and sea surface temperature measurements. The procedure was successfully tested with the limited amount of observational data capable of evaluating such a technique.

The incorporation of stability effects in the meteorological specification makes possible the inclusion of such effects in wave prediction. As a part of this study, the wave spectral growth formulation of the wave prediction model under development at N.Y.U. was generalized to include certain effects of stability in the following way.

1. The effect of stability upon the Miles-Phillips instability mechanism was investigated and it was demonstrated that the existing relation between growth rate and surface shear stress was of sufficient generality to describe the effects of stability on growth rate, provided a stability dependence was incorporated into the calculation of the shear stress from the wind field.

2. The Pierson-Moskowitz fully developed spectral form was effectively generalized to non-neutral conditions by relating the fully developed form to the low level wind profile, instead of to the wind high above the surface. Within the context of the spectral growth formulation, this effect allows stability to further affect the growth rate as well as the maximum height to which a spectral component can grow for a given anemometer level and speed, fetch and duration.

3. Through the application of the surface boundary layer model, the stability modifications were tested by generating model predictions for conditions corresponding to those encountered in observational studies of the dependence of wave generation on air-sea temperature difference. The calculations compared quite favorably with the observational evidence.

4. The effects of stability on wave generation are significant with the extremes of air-sea temperature frequently encountered over large areas of the major oceans. For example, the range of air-sea temperature difference, -7°C to +4°C, produces deviations in significant wave height over neutral conditions that average +20% over the wind speed range 20-40 knots.

In terms of the spectrum, these changes correspond to significant shifts in the spectral peak and larger changes in the spectral density of wave components at frequencies near and below that of the spectral peak.

The objective analysis procedure utilizes ships' wind observations to correct the initial guess wind field, which is standardized for convenience to 19.5 meters above sea level. In the procedure, ship observations are allowed to affect the nearest grid point and are incorporated into the analysis in an order determined by the type of the observation. Observations from ocean station vessels are given highest priority, followed by observations known to represent measurements from anemometers at known elevations and finally the Beaufort estimates. Measured winds are corrected to 19.5 meters on the basis of stability dependent profile forms prior to incorporation into the analysis. Beaufort estimates are standardized to 19.5 meters by the application of a set of Beaufort force–19.5 meters wind speed equivalents that were developed as a part of this study from data collected on the British and Canadian weather ships.

Within the context of the surface boundary layer model derived in this study, the corrected 19.5 wind analysis and the air-sea temperature difference analysis fully define the distribution of wind and temperature and the surface shear stress in the marine surface layer.

Appendix B
Wind Field Analysis Programs Under Development
by Vincent Cardone

The generation of wind fields for the North Pacific hindcasting program requires the application of the following three programs.

DIFNAL

This program produces analyses of the air-sea temperature difference on the portion of the JNWP grid covering the North Pacific Ocean from about 20°N, at 12-hour intervals. Input consists of ship reports from teletype (Monterey line) and NWRC files. The objective analysis technique employed (CRAM-SMOOTH*), requires a realistic initial guess field both on the boundary of and in the interior of the grid. DIFNAL generates an initial guess field by applying CRAM-SMOOTH (assuming a zero initial field) to a composite of ship reports for 3 synoptic observation times centered on the analysis time. The boundary values are computed in the same way by the application of a one-dimensional form of CRAM-SMOOTH, except for the low latitude boundary which is set to a value determined by climatology. The final analysis is generated by applying CRAM-SMOOTH to snyoptic observations only, on the computed initial guess field, with appropriate error checking procedures.

WNDCNV

This program produces analyses of meridional and zonal components of the 19.5 meter wind speed on a JNWP grid expanded to cover the entire No. Pacific Ocean. Input consists of standard JNWP grid analyses of sea level pressure, air temperature, air-sea temperature difference, and Northern Hemisphere grid values of climatological mean monthly meridional and zonal surface winds. The wind components are calculated each 6 hours on the standard grid by the application of a two-layer baroclinic-diabatic planetary boundary layer model to the pressure and temperature data. The computed winds are merged with the appropriate climatological fields on the outer boundary of the standard grid by a ring-scan and component-wise averaging of the computed and climatological winds on the two outer rows of grid points of the standard grid. The program is also capable of providing at each grid point an estimate of the surface shearing stress and the Monin-Obukov scale length.

DWNPCV

This program uses the output of WNDCNV, DIFNAL and ship reports of wind speed and

---

*A combination of conditional relaxation and smoothing.

direction to produce analyses of the 19.5 meter wind speed and direction and surface shearing stress on an expanded JNWP grid but with the grid spacing halved. The first part of the program, then, interpolates the analysis produced by DIFNAL and WNDCNV to the new grid system. The objective analysis scheme employed to incorporate ship reports into the analyses generated by WNDCNV is CRAM-SMOOTH. Ship reports are incorporated into the analysis in a way consistent with the type and suspected accuracy of the observation. All ship wind reports are placed in one of the following four categories:

1. Weather ship on station

2. Wind measured-anemometer height known

3. Wind measured-anemometer height unknown

4. Beaufort estimates

Reports in category 1 are corrected to 19.5 meters through the application of stability dependent profile forms and given highest priority, in the instance that two or more reports affect the same grid point. Reports in category 2 are corrected in the same way and given priority over reports in categories 3 or 4. Reports in category 3 are placed at 19.5 meters and given priority over those in category 4. Category 4 reports are corrected to 19.5 through the application of a new set of Beaufort number-wind speed equivalents. The final analyses of 19.5 meter wind combined with the air-sea temperature difference analysis are used to compute the final analyses of surface shearing stress.

References

1. Pierson, W. J., I. J. Tick and L. Baer. "Computer Based Procedures for Preparing Global Wave Forecasts and Wind Field Analyses Capable of Using Wave Data Obtained by a Spacecraft," *6th Naval Hydrodynamics Symposium*, ARC-136, Office of Naval Research, 1966.

2. Miles, J. W. "On the Generation of Surface Waves by Shear Flow," *J. Fluid Mech.*, Part 1, 3 (1957), 184-204; Part 2, 6 (1959), 568-582; Part 3, 6 (1959), 583-598; Part 4, 13 (1962), 433-448.

3. Phillips, O. M. *The Dynamics of the Upper Ocean*. Cambridge University Press, 1966.

4. Inoue, T. "On the Growth of the Spectrum of a Wind Generated Sea According to a Modified Miles-Phillips Mechanism and the Application to Wave Forecasting," GSL Report TR-67-5, New York University, 1967.

5. Marks, W., T. R. Goodman, W. J. Pierson, Jr., L. J. Tick and I. A. Vassilopoulos. "An Automated System for Optimum Ship Routing," *Trans. Soc. Naval Arch. and Marine Engineers*, 76 (1968), 22-25.

6. Bunting, D. C. and L. I. Moskowitz. "An Evaluation of a Computerized Numerical Wave Prediction Program for the North Atlantic Ocean," Technical Report, U.S. Naval Oceanographic Office (unpublished, but in preparation), 1969.

7. Pierson, W. J. and L. Moskowitz. "A Proposed Spectral Form for Fully Developed Wind Seas Based on the Similarity Theory of S. A. Kitaigarodskii," *J. Geophys. Res.*, 59, No. 29 (1964), 5161-5179.

8. Cardone, V. J. (1969), "Specification of the wind distribution in the marine boundary layer for wave forecasting." N.Y.U. Tech. Report TR-69-1, Dept. of Meteorology and Oceanography, New York University."

16

spacecraft altimetry and ocean geoid

Introduction

Various studies of the application of radar altimeter to geodesy and oceanography have been published in the past few years, e.g., Ewing,[1] Frey et al.,[2] Pierson et al.,[3] and Greenwood et al.[4,5] Greenwood et al. described a relatively simple radar system that appears to be technologically feasible and showed how measurements with such a radar altimeter could provide valuable information for both geodesy and oceanography. The accuracy that appeared to be available at the time of this study was in the ±5 to ±1 meter range. Since the publication of these results a number of important developments have occurred, which will be summarized below.

Memorandum Change 20[6]

The above document, released on March 20, 1969, described opportunities to participate in a Geos-C radar altimeter experiment subject to its anticipated approval. The experiment aimed at an altimeter with a ranging accuracy of ±5 meters and described some of the many results that could be achieved. Even with this accuracy, much could be learned about the geoid well beyond what has been learned by studying the orbit perturbations of various spacecraft. Since then it has become fairly clear that the ±1 meter accuracy is obtainable for Geos-C, and questions as to the choice of orbit arise.

The primary objective of the Geos-C spacecraft is to provide geodetic tracking data that is required to complete the gravimetric and geometric investigations of the National Geodetic Satellite Program. The secondary objective is to demonstrate the feasibility of using a satellite-borne radar altimeter for obtaining information on the geometry of the oceans.

The decision to try a radar altimeter was based on the studies cited above and on a study conducted by Raytheon of the relative advantages and disadvantages of radar and laser altimetry. The radar altimeter won on many counts such as being lighter, capable of seeing through clouds, and at a more advanced state of the art.

NASA Planning

The use of a radar altimeter as a means of studying the earth is well along in the plans of

NASA through the next decade. Low altitude and drag free satellites as well as geodetic satellites with altimeters are under development. Between now and 1980, a succession of spacecraft with altimeters of increasing accuracy, first 1 meter, then 50 cm and 20 cm, and finally 10 cm, are projected. Each spacecraft will be capable of providing useful oceanographic information, and the 10 cm goal, when reached, will be of great value to oceanography.

A Summer Study

A study was held during the summer of 1969 at Williams College in Williamstown, Massachusetts, to review the Earth Physics program of NASA and prepare a scientific document on problems of instrumentation, geodetic measurements and oceanographic applications. This document has been published.

My impressions of some of the results of this study were that the radar altimetry problem achieved additional definition, that the geodesists were able to demonstrate further the advantages of altimetry to geodesy, and that both the benefits to oceanography and the need for strong oceanographic inputs were documented.

Radar altimetry ranging error depends on the ability to measure time to an accuracy of a nanosecond and on getting enough independent measurements of the return signal to be able to average out the effects of sampling variability. A nanosecond capability in timing appears to be technologically feasible now, and in less than a decade, the tenth nanosecond capability appears to be possible. The sampling variability problem can be overcome by a redundant pulse repetition frequency so that many thousands of pulses per second will be transmitted.

The geodesists at the summer study extrapolated what they would be able to obtain by continued application of present techniques and found further substantiation of what Greenwood et al.[5] had concluded namely that the techniques were self-limiting. The definition of the geoid over ocean areas would be markedly improved even with ±5 meter accuracy.

Some recent developments in oceanography will aid in the use of the altimeter data. At Scripps, ways which require only about 20 minutes of computer time to compute the co-tidal and co-height lines for a given tidal frequency over the world ocean have been developed. It should be possible to obtain this kind of solution for all of the major tidal frequencies and to predict the tidal movements of the sea surface in space-time so as to remove these effects along every orbit. Inconsistencies that arise can be used to redefine the tidal predictions so that the tidal effects can finally be completely removed.

Accuracies of 10 to 50 centimeters in ranging make it necessary to consider the effects of wind waves and swell in biasing the range determination. Unpublished work in progress by the writer suggests that these biases can be determined and that adequate wave specification techniques as described in my accompanying paper on wave forecasting will make it possible to remove this bias.

References

1. Ewing, G. C. (ed.). *Oceanography from Space.* Woods Hole Oceanographic Institution, 1965.

2. Frey, E. J., J. V. Harrington and W. S. von Arx. "A Study of Satellite Altimetry for Geophysical and Oceanographic Measurement," *Proc. 16th Congr. Intern. Astronaut. Federation,* 1965.

3. Pierson, W. J., B. B. Scheps and D. S. Simonett. "Some Applications of Radar Return Data to the Study of Terrestrial and Ocean Phenomena," *Proc. 3rd Goddard Mem. Symp. Sci. Expts. Manned Orbital Flights,* 1965, 87-137.

4. Greenwood, J. A., A. Nathan, G. Neumann, W. J. Pierson, J. F. Jackson and T. E. Pease. "Radar Altimetry from a Spacecraft and Its Potential Applications to Geodesy," *Remote Sensing of Environment,* 1, No. 1 (1969), 59-70.

5. _____. "Oceanographic Applications of Radar Altimetry from a Spacecraft," *Remote Sensing of Environment,* 1, No. p (1969), 71-80.

6. Memorandum Change 20, Opportunities for
    Participation in Space Flight Investigations
    Subject Geos-C Radar Altimeter Experiment

dated March 20, 1969, (with a covering letter
by John E. Naugle, associate administrator for
Space Science and Applications).

other volumes
in the

topics
in
ocean
engineering
series

topics in ocean engineering

volume 1

Topics in Ocean Engineering, Vol. 1 was published in 1969 by Gulf Publishing Company. The Editor of Volume 1 is also Charles L. Bretschneider, Chairman of the Department of Ocean Engineering at the University of Hawaii. He is also the editor of the three volumes of this series which will be published in the future.

The contributors to Volume 1 of this series are Manley St. Denis, Basil W. Wilson, Milton S. Plesset, Robert Taggart, H. William Gillen, Giulio Venezian, Daniel K. Ai, and F. Gerritsen. The Contents of *Topics in Ocean Engineering,* Vol. 1 is reprinted on the following pages.

contents

volumes 3, 4 and 5 of this series

The topics treated in volumes to follow in this series are based on lectures during the annual seminars at the University of Hawaii Sea Grant Program sponsored by the National Science Foundation.

Volume 1 of *Topics in Ocean Engineering* was published in 1969 following the first seminars in this program.

Among the authors to be included in Volumes 3, 4, and 5 to be published in 1971, 1972, and 1973 are Dr. Richard Paul Shaw, who is a contributor to this volume; Dr. William Van Dorn, Scripps Institute of Oceanography at the University of California and Dr. Taivo Laevestu at Fleet Numerical Weather Central.

Topics to be published in future volumes include material on experimental investigation of structures for deep sea application, long period harbor oscillations, tsunami engineering, a comparative review of oceanographic processes in coastal waters, the application of a hydrodynamical-numerical (HN) model of Walter Hansen for prediction of dispersion of pollution, and the classification and forecasting of near-surface ocean thermal structure.